Y0-BYK-810

CONFORMATIONAL ANALYSIS

Scope and Present Limitations

ORGANIC CHEMISTRY

A SERIES OF MONOGRAPHS

ALFRED T. BLOMQUIST, *Editor*

Department of Chemistry, Cornell University, Ithaca, New York

1. Wolfgang Kirmse. CARBENE CHEMISTRY, 1964; 2nd Edition, *In preparation*

2. Brandes H. Smith. BRIDGED AROMATIC COMPOUNDS, 1964

3. Michael Hanack. CONFORMATION THEORY, 1965

4. Donald J. Cram. FUNDAMENTAL OF CARBANION CHEMISTRY, 1965

5. Kenneth B. Wiberg (Editor). OXIDATION IN ORGANIC CHEMISTRY, PART A, 1965; PART B, *In preparation*

6. R. F. Hudson. STRUCTURE AND MECHANISM IN ORGANO-PHOSPHORUS CHEMISTRY, 1965

7. A. William Johnson. YLID CHEMISTRY, 1966

8. Jan Hamer (Editor). 1,4-CYCLOADDITION REACTIONS, 1967

9. Henri Ulrich. CYCLOADDITION REACTIONS OF HETEROCUMULENES, 1967

10. M. P. Cava and M. J. Mitchell. CYCLOBUTADIENE AND RELATED COMPOUNDS, 1967

11. Reinhard W. Hoffman. DEHYDROBENZENE AND CYCLOALKYNES, 1967

12. Stanley R. Sandler and Wolf Karo. ORGANIC FUNCTIONAL GROUP PREPARATIONS, VOLUME I, 1968; VOLUME II, *In preparation*

13. Robert J. Cotter and Markus Matzner. RING-FORMING POLYMERIZATIONS, PART A, 1969; PART B, *In preparation*

14. R. H. DeWolfe. CARBOXYLIC ORTHO ACID DERIVATIVES, 1970

15. R. Foster. ORGANIC CHARGE-TRANSFER COMPLEXES, 1969

16. James P. Snyder (Editor). NONBENZENOID AROMATICS, I, 1969

17. C. H. Rochester. ACIDITY FUNCTIONS, 1970

18. Richard J. Sundberg. THE CHEMISTRY OF INDOLES, 1970

19. A. R. Katritzky and J. M. Lagowski. CHEMISTRY OF THE HETEROCYCLIC N-OXIDES, 1970

20. Ivar Ugi (Editor). ISONITRILE CHEMISTRY, 1970

21. G. Chiurdoglu (Editor). CONFORMATIONAL ANALYSIS, 1971

In preparation

Gottfried Schill. CATENANES, ROTAXANES, AND KNOTS

International Symposium on Conformational
Analysis, Brussels, 1969.

CONFORMATIONAL ANALYSIS

Scope and Present Limitations

Papers presented at the
Brussels International Symposium
(September 1969)

Edited by

G. Chiurdoglu

Société Chimique de Belgique
Brussels, Belgium

QD481
A1
I618c
1969
(1971)

1971

ACADEMIC PRESS
New York and London

263646

COPYRIGHT © 1971, BY ACADEMIC PRESS, INC.
ALL RIGHTS RESERVED
NO PART OF THIS BOOK MAY BE REPRODUCED IN ANY FORM,
BY PHOTOSTAT, MICROFILM, RETRIEVAL SYSTEM, OR ANY
OTHER MEANS, WITHOUT WRITTEN PERMISSION FROM
THE PUBLISHERS.

ACADEMIC PRESS, INC.
111 Fifth Avenue, New York, New York 10003

United Kingdom Edition published by
ACADEMIC PRESS, INC. (LONDON) LTD.
Berkeley Square House, London W1X 6BA

LIBRARY OF CONGRESS CATALOG CARD NUMBER: 72-125945

PRINTED IN THE UNITED STATES OF AMERICA

Contents

List of Contributors

Numbers in parentheses indicate the pages on which authors' contributions begin.

M. AGLIETTO, Istituto di Chimica Organica Industriale dell'Università, Pisa, Italy (203)

C. ALTONA, Department of Chemistry, The University, Leiden, The Netherlands (1)

F. A. L. ANET, Department of Chemistry, University of California, Los Angeles, California (15)

M. J. O. ANTEUNIS, Department of Organic Chemistry, State University of Ghent, Ghent, Belgium (31)

K. VON BREDOW, Institut für Elektrowerkstoffe der Fraunhofer-Gesellschaft Freiburg and the Chemisches Laboratorium der Universität, Freiburg, Germany (51)*

ROBERT BUCOURT, Centre de Recherches Roussel-Uclaf, Romainville, France (59)

MEHRU J. COOPER, Department of Chemistry, Laval University, Quebec, Canada (73)

HUGH FELKIN, Institut de Chimie des Substances Naturelles, C.N.R.S., Gif-sur-Yvette, France (63)

GABOR FODOR, Department of Chemistry, West Virginia University, Morgantown, West Virginia (73)

DANIEL FREHEL, Department of Chemistry, Laval University, Quebec, Canada (73)

* Present address: Geigy AG, Basel, Switzerland.

H. Friebolin, Institut für Elektrowerkstoffe der Fraunhofer-Gesellschaft Freiburg and the Chemisches Laboratorium der Universität, Freiburg, Germany (51)*

Edgar W. Garbisch, Jr., Department of Chemistry, University of Minnesota, Minneapolis, Minnesota (93)

Bruce L. Hawkins, Department of Chemistry, University of Minnesota, Minneapolis, Minnesota (93)

V. T. Ivanov, Institute for Chemistry of Natural Products, USSR Academy of Sciences, Moscow, USSR (111)

S. Kabuss, Institute für Elektrowerkstoffe der Fraunhofer-Gesellschaft Freiburg and the Chemisches Laboratorium der Universität, Freiburg, Germany (51)

Jean-Marie Lehn, Institut de Chimie, Strasbourg, France (129)

P. L. Luisi, Technisch-Chemisches Laboratorium, Eidgenössische Technische Hochschule, Zurich, Switzerland (203)

Robert E. Lyle, Department of Chemistry, University of New Hampshire, Durham, New Hampshire (157)

Kenneth D. MacKay, Department of Chemistry, University of Minnesota, Minneapolis, Minnesota (93)

James McKenna, Chemistry Department, The University, Sheffield, England (165)

Nagabhushanam Mandava, United States Department of Agriculture, Beltsville, Maryland (73)

Jeremy I. Musher, Belfer Graduate School of Science, Yeshiva University, New York, New York (177)

K. Onodera, Laboratory of Biological Chemistry, Department of Agricultural Chemistry, Kyoto University, Kyoto, Japan (191)

Yu. A. Ovchinnikov, Institute for Chemistry of Natural Products, USSR Academy of Sciences, Moscow, USSR (111)

P. Pino, Technisch-Chemisches Laboratorium, Eidgenössische Technische Hochschule, Zurich, Switzerland (203)

S. Pucci, Centro di Chimica delle Macromolecole, Pisa, Italy (203)

J. Reisse, Faculté des Sciences, Université Libre de Bruxelles, Brussels, Belgium (219)

Gottfried Schill, Chemisches Laboratorium der Universität, Freiburg, Germany (229)

Paul von Ragué Schleyer, Department of Chemistry, Princeton University Princeton, New Jersey (241)

Robert D. Stolow, Department of Chemistry, Tufts University, Medford, Massachusetts (251)

* Present address: Institut für Makromolekulare Chemie, Universität Freiburg, Freiburg, Germany.

JOHN J. THOMAS, Department of Chemistry, University of New Hampshire, Durham, New Hampshire (157)

DAVID A. WALSH, Department of Chemistry, University of New Hampshire, Durham, New Hampshire (157)

H. H. WESTEN, Organisch-chemisches Laboratorium, Eidgenössische Technische Hochschule, Zurich, Switzerland (259)

Preface

This work contains the full text of twenty papers read at the International Symposium on Conformational Analysis, which took place in Brussels, September 1969. These papers were selected by the International Scientific Committee. which was comprised of Professor G. Chiurdoglu (Chairman), J. Reisse (Secretary), D. H. R. Barton (London), A. Dreiding (Zurich), E. Havinga (Leiden), A. Luttringshaus (Freiburg i.B.), G. Ourisson (Strasbourg), V. Prelog (Zurich), R. H. Martin (Brussels), J. Nasielski (Brussels), and J. van Binst (Brussels).

The treated subjects cover the present long-range research fields in conformational analysis: theoretical aspects and applications such as those concerning the aliphatic and especially the cyclic series. These problems are studied by specialists who have viewed this subject with a wider applicability New results were thoroughly and critically discussed. The references are very useful in each case.

First, this book attempts to review the entire subject, to point out the areas in need of further research, and also to detect difficulties which have not yet been solved. The proceedings will help the reader determine the present status and the prospects for future developments in this continuously expanding field of chemistry. This volume should be kept as a reference book in every stereochemistry laboratory.

The success of the Brussels Symposium was assured by the high standard of scientific contributions; the Scientific Committee, the authors' plenary lectures (published by Butterworths, London), the quality of the contributors' papers and also of all participants attending the Symposium, and the sponsorship of the I.U.P.A.C. all added to this success.

The valuable contributions of the Organizing Committee and, in particular, of its Executive Secretary, Richard C. Smekens, whose devotion, experience, and efficiency ensured materially the success of the Congress and permitted the publication of these proceedings, must be acknowledged.

Geometry of Five-Membered Rings

C. Altona
Department of Chemistry
The University
*Leiden, The Netherlands**

The conformational analysis of nonplanar five-membered ring compounds is considerably more complicated than that of chair-shaped six-membered rings. The latter, thanks to their relative rigidity (high resistance to deformation by bulky groups) are easily characterized by the orientation of substituents, equatorial or axial, and by a well-defined geometry for each conformation. Recent computer calculations of substituted cyclohexanes in conjunction with x-ray determinations taken from the literature have shown (1) that common e and a substituents (chlorine, methyl, equatorial t-butyl) do induce definite, but usually minor, deformations of the chair form.

In contrast, the five-membered rings are pseudorotational systems in the liquid and gas phase. In cyclopentane the puckering rotates around the ring without intervention of substantial potential-energy barriers (2–4). On introducing one or more substituents or hetero atoms into the ring, a potential-energy barrier (or barriers) restricting pseudorotation may occur. "Classical" conformational analysis, presupposing a defined geometry, *is possible only if the properties of the substituent(s) are such that the induced barrier is sufficiently high compared to the pseudorotational energy levels.* Examples where this condition is not fulfilled are perhaps chlorocyclopentane and the diequatorial form of *trans*-1,2-dihalocyclopentanes in solution at room temperature (5–7), as evidenced by their infrared and Raman spectra.

Various terminologies to describe the conformation of five-membered rings on the basis of the two forms that contain elements of symmetry are currently

* Part of the investigations reported herein were carried out at the Department of Chemistry, Case Western Reserve University, Cleveland, Ohio, during leave of absence from the University of Leiden.

1

used (8, 9). We have the C_s or envelope and the C_2 or half-chair form, which have been termed "basic conformations" (5). Taking the energy differences between axial and equatorial position into account, the ring may assume eleven *energetically different* basic conformations for a monosubstituted cyclopentane, five C_2 and six C_s forms (see reference 5 for a more complete discussion). However, a large amount of evidence has recently accumulated, mainly from x-ray structure studies (10, 11), to the effect that in the vast majority of cases none of these basic forms actually represents a minimum of energy. *The true geometry is usually somewhere "in between," neither C_2 nor C_s.* Only in simple cases, viz. where a single, conically symmetrical, substituent is present or where such substituents are symmetrically placed, one may safely consider the

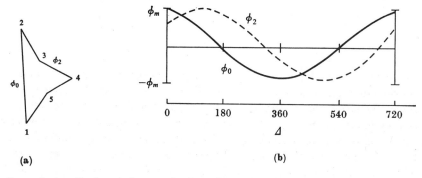

(a)   (b)

Figure 1   Amplitude and phase angle of pseudorotation of a five-membered ring.

appropriate basic conformation or conformations to represent the molecule in its ground state.

In what follows, the discussion of conformational aspects of some five-membered ring compounds will be based mainly on observed (diffraction methods) and calculated torsional angles, since these have the virtue of being highly sensitive to changes in the pseudorotation parameter. Hence we review briefly the interrelationship between the five torsional angles in a general cyclopentane ring (10, 12), which relationship can be used to define both the exact "point" (Δ) on the pseudorotation pathway as well as the exact degree or amplitude of puckering (ϕ_m).

Consider the half-chair form (Fig. 1), where the torsional angle about the 1–2 bond ϕ_0 is taken equal to ϕ_m with positive sign. Along the pseudorotation itinerary ϕ_0 decreases (solid line of Fig. 1b) until at $\Delta = 360°$ the mirror image of the original structure is obtained: $\phi_0 = -\phi_m$. A rotation over $\Delta = 720°$ restores the original conformation, Δ *is called the phase angle of pseudorotation.* Each torsional angle (see ϕ_2, Fig. 1) follows an identical pattern with a certain lag in Δ. Mathematically, this relation between angles can be expressed in several ways. A convenient formulation is given in equation (1).

$$\phi_j = \phi_m \left(\cos \frac{1}{2} \Delta - \frac{4}{5} \pi j \right) \qquad (j = 0, 1, 2, 3, 4) \tag{1}$$

If Δ is taken as zero for an arbitrary C_2 form, C_2 forms arise at $\Delta = 0$, 72, 144°, ..., and C_s forms at $\Delta = 36, 108, 180°, ...$ (Fig. 2).

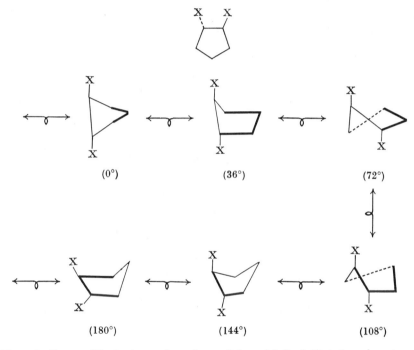

(0°) (36°) (72°)

(180°) (144°) (108°)

Figure 2 Six possible basic conformations of *trans*-1,2-disubstituted cyclopentanes representing one-quarter of the pseudorotation circuit and the corresponding phase angles Δ.

The conformational analysis of ring D in about 35 steroids (10), where equation (1) was used as the main analytical tool, has shown that in only a few molecules, ring D occurs as $C_s(14)$ envelope ($\Delta \simeq -36°$), as $C_s(13)$ envelope ($\Delta \simeq +36°$) or as half-chair ($\Delta \simeq 0°$), respectively. The remaining compounds analyzed have ring D occurring in forms intermediate between half-chair and envelope. Some characteristic figures are androsterone, $\phi_m = 45.2°$, $\Delta = -7.6°$; and $5\alpha,17\beta$-cholestanes, $\phi_m = 46-50°$, $\Delta = 8-17°$. A similar analysis of a variety of hetero five-membered rings is given in reference 11.

I. CONFORMATIONAL ANALYSIS IN SOLUTION

A concerted attack on conformational problems in mono-, di- and trihalogeno-cyclopentanes and related compounds by means of electric moments, nuclear magnetic resonance (nmr) and vibration spectroscopy yielded a consistent picture of their properties and showed that in favorable cases "classical" analysis (determination of equilibrium constants, solvent effects and so on) can be carried out with success (*5–7, 13–16*) (see Table I).

Table I Approaches to Conformational Analysis of Five-Membered Rings

State	Method	Examples
I Solid	x-Ray crystallography	Ring D in steroids (*10*); furanoid systems (*10a*); hetero rings (O,N,P,S) (*11*); norbornanes (*29*)
II Solution	Nuclear magnetic resonance (variation of J_{vic} with torsional angle); dipole moments); (μ^2/J method); infrared/Raman (carbon–halogen stretching frequencies)	Monohalogenocyclopentanes (*7*); *trans*-1,2-dihalogenocyclopentanes (*5*); 1,1,2-trihalogenocyclopentanes (*10b*); *trans*-1,2-dihalogenoindans (*15*); 3-halogenotetrahydrofurans (*14*); *trans*-3,4-dihalogenotetrahydrofurans (*14*); 2-substituted-1,3-dioxolanes (*17*)
III Gas phase	Electron diffraction (vibration spectroscopy, microwave)	Cyclopentane (*17a*); tetrahydrofuran (*17b*); norbornane (*30*)
IV Theoretical	Valence-force-field calculations (Westheimer method)	Ring D in steroids; norbornanes

This work led to the following generalizations:

(1) The substituents (Cl and Br) prefer to occupy the puckered part of the ring.

(2) In the *trans-vic*-dihalogenides of cyclopentane, indan and tetrahydro-furan, the chlorides are isogeometrical with the bromides.

(3) An equilibrium aa $\rightleftharpoons$ ee exists.

(4) The aa form predominates in all cases investigated, even in highly polar solvents. The dibromides tend to exist almost exclusively (>90%) in the diaxial form.

It might be mentioned in passing that the well-known holding effect of a *t*-butyl group in cyclohexane chemistry cannot be applied to cyclopentanes. Consideration of models shows readily that, for example, in *trans*-1,2-dibromo-4-*t*-butylcyclopentane, the diaxial and diequatorial forms may equilibrate via the low-barrier pseudorotation pathway, while the bulky t-butyl group remains equatorial. In effect, a dipole-moment study (*16*) of the compound mentioned showed that two forms, aa and ee, are in dynamic equilibrium with the diaxial form predominating.

The effect of side-chain orientation on ring geometry of 2-alkoxy-1,3-di-oxolanes was recently studied by means of dipole moments and AA'BB' coupling constants (*17*). The results were interpreted on the basis of a gauche–anti equilibrium (Fig. 3). For R = Me, the anti form seems to be preferred over the gauche form. In contrast, for R = t-butyl, steric strain overcomes stereoelectronic factors (*11, 18*) and the t-Bu derivative prefers to exist mainly in the gauche form. The most interesting feature of this shift in equilibrium with the size of the 2 substituents is the accompanying change of J_{cis}. This was rationalized on the basis of a change in geometry (pseudorotation), the torsional angle in the H_2C–CH_2 fragment being smaller in the gauche than in the anti conformation.

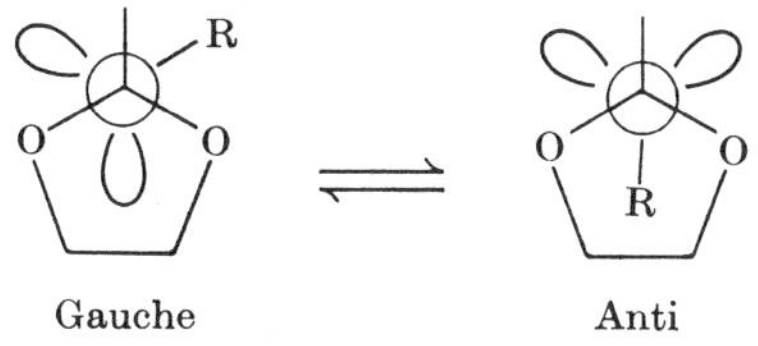

Gauche   Anti

Figure 3   Anti and gauche conformers of 2-alkoxy-1,3-dioxolanes.

II. VALENCE-FORCE CALCULATIONS

One of the main reasons for the relatively slow pace of conformational analysis of five-membered rings is lack of reliable information concerning the shape of the potential barriers separating the conformers and the effect of various substituents and their orientation on the ground-state geometry. It seems to us that further advances require a quantitative understanding of the energetical and geometrical factors involved. Particularly promising are large-scale computer calculations (Westheimer method) of series of compounds, utilizing full relaxation of internal coordinates (*19*) during the energy-minimization procedure (*4, 19–22*). Before one can have confidence in the reliability of this approach, it should be shown that the force field and its parameters are able to reproduce known conformational energies and geometries. The remainder of this paper is devoted to the correlation between calculated and observed structural details and pseudorotation in some compounds that contain one or more five-membered rings. Lack of pertinent data on simple substituted cyclo pentanes dictated the present choice of more complicated systems, viz. steroids and norbornanes, for which much experimental information is available.

The force-field equations and parameters used are based on those developed by Allinger *et al.* (*21*), but incorporate a number of innovations (see Table II). Details will be presented elsewhere (*1, 23, 29*).

Table II   Valence-Force-Field Calculations (Utah Program (*22*))

Potential energy is minimum with respect to all internal coordinates (*29*):

$$V = V_{stretch} + V_{bend} + V_{torsion} + V_{nonbonded} + V_{1,3} + V_{coulomb}$$

$V_{stretch}$: $\frac{1}{2}k_s(1 - 1_0)^2 + k'(1 - 1_0)$

V_{bend}: Allinger *et al.* (*21*)

$V_{torsion}$: Allinger *et al.* (*21*)

$V_{nonbonded}$: Allinger *et al.* (*21*) (Hill-type exponential function)

$V_{1,3}$: Hill-type 6/12 function (80–85% of "normal" nonbonded radius)

$V_{coulomb}$: Between fractional electronic charges located on the nuclei

Iterative method, in final cycle:

maximum energy change < 0.001 kcal/mole

maximum bond distance changes ≪ 0.001 Å

maximum bond angle changes < 0.01°

maximum number of atoms: currently 65

maximum number of interactions: currently 1500

III. CALCULATION OF STEROIDS

The compounds chosen are androsterone (I), $5\alpha,17\beta$-pregnane-$3\beta,20\beta$-diol (II) and the corresponding 20α-diol (III) (Fig. 4). Four side-chain rotamers of each epimer II and III were examined in an effort to explain the coupling constants in these molecules (*24*). The rotamers differed with respect to the distribution of groups about the C_{17}–C_{20} bond and to the position of the hydroxylic hydrogen atom.

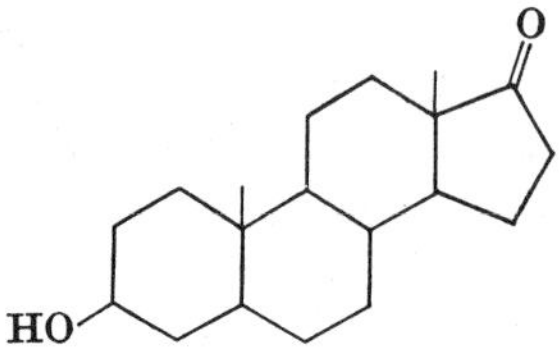

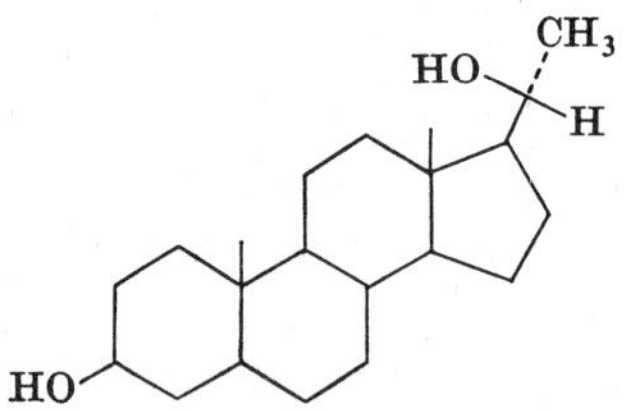

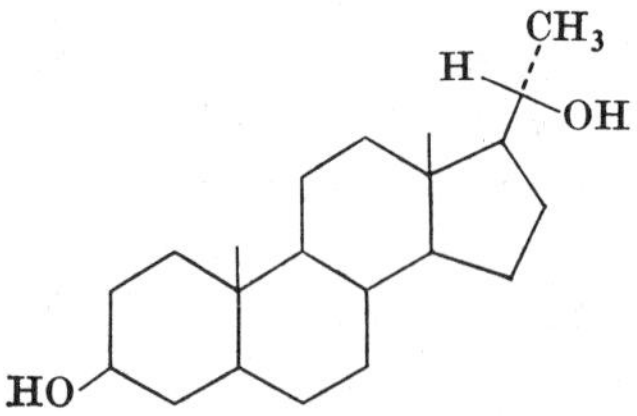

Figure 4   Calculated structures: 5α-steroids.

Table III   Torsional and Valency Angles of Androsterone[a]

	Torsional angles (°)				Bond angles (°)		
	Exp	Calc	|Diff|		Exp	Calc	|Diff|
ϕ_0	44.0	43.6	0.4	17–13–14	99.2	99.8	0.6
ϕ_1	−34.3	−34.2	0.1	13–14–15	104.3	104.5	0.2
ϕ_2	11.5	12.4	0.9	14–15–16	102.6	103.2	0.6
ϕ_3	16.6	15.0	1.6	15–16–17	106.0	105.5	0.5
ϕ_4	−38.6	−36.9	1.7	16–17–13	107.8	108.0	0.2
		$\sigma = 0.8$				$\sigma = 0.4$	

[a] Experimental (25) and calculated.

The experimental and calculated features of ring D in I are shown in Table III. The 17-keto group causes the torsional barriers about C_{13}–C_{17} and C_{16}–C_{17} to be lower than those about C_{14}–C_{15} and C_{15}–C_{16}; as a consequence, the conformation will tend toward negative $\varDelta$ values (Fig. 5). Indeed, the few rings D known today, that have negative phase angles $\varDelta$ all carry a

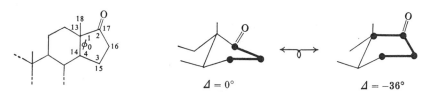

$\varDelta = 0°$ $\varDelta = -36°$

Figure 5   Ring D of androsterone showing experimental and calculated phase angle of pseudorotation ($\varDelta_{\text{exp}} = -7.6°$, $\varDelta_{\text{calc}} = -4.2°$).

17-keto substituent (10). It is reason for satisfaction to find that the calculated torsional and valency angles (and the bond distances as well, not shown) agree with the x-ray determination *within the limits of error of the latter.* The calculated $\varDelta$ (−4.2°) compares well with the value from the experimental coordinates (−7.6°).*

* Recent calculations (23) show that the "strain energy" changes little during pseudo-rotation of ring D of androsterone. Putting the minimum energy form ($\varDelta = -7.6°$) as zero, we find for the C_{13} envelope form ($\varDelta = +36°$) a relative energy increase of 0.8 kcal/mole; for the C_{14} envelope an increase of only 0.26 kcal/mole. In other words, the potential energy well of a 17-keto substituted ring D is quite shallow and relatively small conformation transmission effects are expected to reveal themselves in the phase angle of pseudorotation. This is indeed found to be the case. Ring D in estrone (aromatic ring A) has a phase angle $\varDelta = -40°$ (see reference 10, Fig. 9b, for further details).

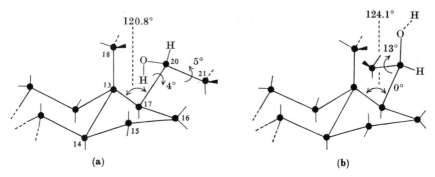

Figure 6 Calculated structure: C/D rings of 5α,17β-pregnane-3β,20β-diol (II) showing two conformations of the side-chain: (a) 20β-G(sc⁺), $\phi_m = 45.3°$, $\Delta = 14.6°$; (b) 20β-H(ap), $\phi_m = 44.2°$, $\Delta = 13.2°$ (*24*).

Figures 6 and 7 show some calculated structural features of the side-chain rotamers of the pregnanediols II and III (*24*), and the calculated Δ values in this series range from +13.1 to +14.6°. In other words, *the conformation of ring D in compounds of this type appears to be hardly affected by the conformation of the side-chain.** No x-ray structure determinations are available for

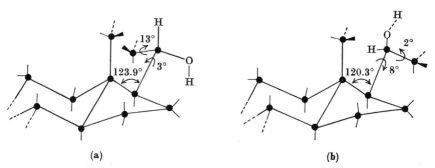

Figure 7 Calculated structure: C/D rings of 5α,17β-pregnane-3β,20α-diol (III) showing two conformations of the side-chain: (a) 20α-E(sc⁺), $\phi_m = 44.4°$, $\Delta = 13.1°$; (b) 20α-F(ap), $\phi_m = 44.6°$, $\Delta = 14.4°$ (*24*).

compounds II and III and we must be content to compare the calculations with analogous molecules, e.g., 2α,3β-dibromo-5α-cholestane (*26*) (Table IV). Again, the comparison is highly encouraging. Figure 8 shows that the calculated pseudorotation angle Δ lies well within the range recorded from comparable structures. Moreover, the calculations fully corroborate earlier

* Even if the side-chain is subjected to considerable nonbonded strain, witness the fact that the total strain energy of II-20β H(ap) is ~2.4 kcal/mole greater than that of the II-20β G(sc⁺) form (Fig. 6).

Table IV   Experimental (*26*) (x-Ray) Torsional and Valency Angles (Ring D) of 2α,3β-Dibromo-5α-cholestane and Calculated Values in II, Side-Chain Conformation G(sc$^+$) (see Fig. 6 and reference 24)

	Torsional angles (°)				Bond angles (°)		
	Exp	Calc	|Diff|		Exp	Calc	|Diff|
ϕ_0	46.8	45.2	1.6	17–13–14	100.6	100.8	0.2
ϕ_1	−40.4	−39.0	1.4	13–14–15	104.9	104.5	0.4
ϕ_2	20.6	19.5	1.1	14–15–16	102.9	103.9	1.0
ϕ_3	7.6	8.0	0.4	15–16–17	106.9	107.2	0.3
ϕ_4	−33.6	−33.1	0.5	16–17–13	103.2	103.4	0.2
	$\sigma = 2.6$				$\sigma = 1.4$		

conclusions (*10*) that the true energy minimum of 17β-carbon-substituted rings D (5α-androstane skeleton) seems to be near $\Delta = 15°$, and does not correspond to C_2 or C_s symmetry.

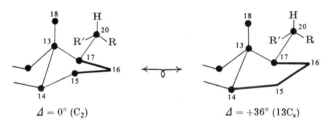

$\Delta = 0°$ (C_2) $\Delta = +36°$ (13C$_s$)

Figure 8   Ring D conformation in 17β-C$_{20}$(sp^3) substituted 5α-steroids—experimental (x-ray): $\Delta = +8 − +17°$, $\Delta_{av} = 13°$; calculated (II and III): $\Delta = +13.1 − +14.6°$.

It should be mentioned here that the distortion of the bond angles in ring D (C–C–C, H–C–C, H–C–H), from the tetrahedral value, is considerable; consequently, the usual assumption made in correlating torsional angles between C–H bonds to ring conformation, that of trigonal symmetry along the projection axis, does not hold at all well (Fig. 9). Much effort has been spent in the past to determine ring D conformations from nmr coupling constants involving 16- and 17-protons. A set of hydrogen–hydrogen torsional angles for the three basic conformations has been obtained by measurement from Dreiding models (*27*) and used on several occasions (*27, 28*). The present force-field calculations do not confirm the validity of this set as far as 17β-substituted steroids are concerned (see Fig. 9 and compare the table in reference 27). Deviations up to an estimated 13° in torsional angle occur for $\Delta = 14°$.*

* Interpolation between values of Cross and Crabbé (*27*) for $\Delta = 0°$ and $\Delta = 36°$ being necessary.

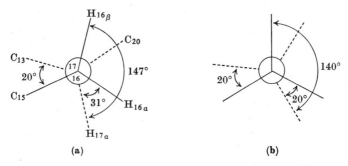

Figure 9   Newman projections along C_{16}–C_{17}, showing hydrogen–hydrogen torsional
angles: (a) calculated structure of II, conformation 20β-G(sc$^+$), (b) same if the
assumption of "trigonal symmetry" were valid.

IV. CALCULATION OF NORBORNANES

The prime motivation for undertaking a theoretical investigation of substituted
norbornanes (viewed as two interlocked five-membered rings) was the interesting discovery (29) that the torsional angles in a number of these systems
(investigated by x-ray methods) did not confirm the C_{2v} symmetry demanded
by a "rigid" cage. It appeared likely that the norbornane system relieves local
steric crowding by a certain twisting of the entire skeleton, i.e., by pseudorotation of *both* five-membered rings. Available x-ray data were insufficient
for a quantitative understanding of the phenomenon, mainly because several
key structures have not been investigated.

Table V   Torsional Angles in Norbornane^a

	Exp (°)	Calc (°)	|Diff| (°)
$a = c$	34.6	35.8	1.2
b	0	0	0
$d = e$	55.3	56.5	1.2
$f = g$	71.8	70.5	1.3

^{a} Experimental (electron diffraction (30)) and calculated. See Fig. 10.

First turning our attention to the structure of the parent compound (Fig. 10
and Table V), it is seen that the calculated geometry, which automatically
converged to the C_{2v} symmetry, closely resembles the geometry found by
electron-diffraction methods (30).

More interesting are the calculated results of 1- and 2-(endo and exo)-substituted norbornanes. *Two modes of twist occur:*

(1) A synchronous pseudorotation of both five-membered rings in the same direction (though not necessarily with the same phase angle *when viewed in projection along* $C_1 \ldots C_4$. This *"synchro"*-twist (Fig. 11) is generally found when substituents (exo or endo) are present on C_2 and/or C_3.

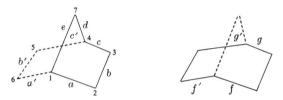

Figure 10   Definition of torsional angles in norbornane skeleton.

(2) A movement whereby the rings twist in opposite direction when viewed along $C_1 \ldots C_4$, which is designated as *"contra"*-twist (Fig. 11).

The latter mode seems to occur mainly in camphanes, camphors and other compounds that are sterically compressed by a bulky substituent on C_1; its magnitude is rather smaller than that of the *synchro*-twist.

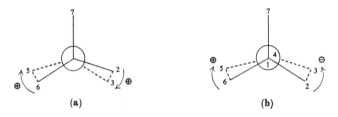

(a)   (b)

Figure 11   Two possible modes of twist in substituted norbornanes: (a) *synchro*-twist, $|a| > |c|, |a'| < |c'|$ (+, +); (b) *contra*-twist, $|a| < |c|, |a'| < |c'|$ (−, +).

Some quantitative results are shown in Table VI. Note that the direction and magnitude of twist most easily follow from a consideration of the torsional angles a and c (Fig. 10). In the case of strict C_{2v} symmetry (norbornane) $a = c \cong 34$–$36°$. A *synchro*-(+,+)-twist is characterized by $a > c$ and $a' < c'$ (disregarding the sign of the torsional angles themselves). Differences between a and c up to 6–9° commonly occur. It is seen (TableVI) that a 2-endo-hydroxyl (IV) and a 3-exo-methyl (V) group both give rise to a clockwise *synchro* (+,+) pseudorotation of the five-membered rings, the methyl group having the greater effect. In combination (2-endo-OH-3-exo-Me, VI) the calculated effect

Table VI   Examples of Torsional Angles in "Twisted" Norbornanes[a]

	Compound	a (°)	a' (°)	b (°)	b' (°)	c (°)	c' (°)
Calc	IV 2-endo-OH	37	36	1	0	35	36
	V 3-exo-Me	39	33	4	4	32	39
	VI 2-endo-OH- 3-exo-Me	40	33	5	4	31	39
Exp (31)	VII 2-endo-OH-3- exo-CH$_2$N(Me)ϕ	44	31	9	5	29	39
Calc	VIII Camphane	34.7	34.7	0.4	0.4	35.6	35.6
Exp (32)	IX 1,1-Biapo- camphane	33.4	32.9	1.2	1.7	36.7	37.1

[a] Experimental (x-ray) and calculated structures.

is enhanced;* however, it is still less than that found in the analogous 3-(N-benzyl-N-methylaminomethyl)-2-norbornanol (VII) for which compound experimental torsional angles were obtained from the known atomic coordinates in the crystal (31) (Table VI). In this compound, the difference between angles a and c (a' and c') are amazingly large, 15 and 8°, respectively. Of course, the 3-exo side group in VII is more bulky than the simple methyl in VI, which fact may be responsible for the increased twist.

V. CONCLUSION

Present-day computer calculations of molecular structures and energies are becoming a highly useful and almost indispensable tool in the conformational analysis of pseudorotating (flexible) systems. So far, the physical model used undoubtedly represents only a fairly crude approximation, but the results presented herein show that this approach is capable of yielding trustworthy geometrical details of large molecules, up to the size of steroids, in a relatively short time.†

ACKNOWLEDGMENTS

The author is indebted to Professors M. Sundaralingam and H. Hirschmann for their generous gifts of computer time. Dr. H.R. Buys kindly assisted in preparing the manuscript for publication.

* A similar enhancement appeared from the calculation of 2-exo-5-exo-dimethylnorbornane (Altona, unpublished result).
† The machine time used in a complete energy minimization of a typical norbornane or camphane (19–30 atoms) was of the order of 30–60 sec. The time increased sharply with the number of atoms in the structure. Steroids, containing 59 atoms, required 5–15 minutes to converge completely on a Univac 1108 computer.

REFERENCES

1. C. Altona and M. Sundaralingam, *Tetrahedron* **26**, 925 (1970).
2. K.S. Pitzer and W.E. Donath, *J. Am. Chem. Soc.* **81**, 3213 (1959) and further references cited therein.
3. J.B. Hendrickson, *J. Am. Chem. Soc.* **83**, 4537 (1961); **84**, 3355 (1962).
4. S. Lifson and A. Warshel, *J. Chem. Phys.* **49**, 5116 (1968).
5. C. Altona, H.R. Buys and E. Havinga, *Rec. Trav. Chim.* **85**, 973 (1966).
6. C. Altona, H.R. Buys and E. Havinga, *Rec. Trav. Chim.* **85**, 983 (1966); H.R. Buys, C. Altona and E. Havinga, *Rec. Trav. Chim.* **85**, 998 (1966).
7. H.R. Buys, C. Altona and E. Havinga, *Rec. Trav. Chim.* **87**, 53 (1968).
8. F.V. Brutcher, T. Roberts, S.J. Barr and N. Pearson, *J. Am. Chem. Soc.* **81**, 4915 (1959).
9. L.D. Hall, *Chem. Ind.* **1963**, 950.
10. C. Altona, H.J. Geise and C. Romers, *Tetrahedron* **24**, 13 (1968); H.R. Buys and H.J. Geise, unpublished data.
10a. C. Altona, unpublished data.
10b. H.R. Buys, C. Altona and E. Havinga, *Rec. Trav. Chim.* **86**, 1007 (1967).
11. C. Romers, C. Altona, H.R. Buys and E. Havinga, *in* "Topics in Stereochemistry," Vol. 4, E.L. Eliel and N.L. Allinger, eds., Wiley (Interscience), New York, 1969.
12. H.J. Geise, C. Altona and C. Romers, *Tetrahedron Letters* **1967**, 1383.
13. C. Altona, H.R. Buys, H.J. Hageman and E. Havinga, *Tetrahedron* **23**, 2265 (1967).
14. H.R. Buys, C. Altona and E. Havinga, *Tetrahedron* **24**, 3019 (1968).
15. H.R. Buys and E. Havinga, *Tetrahedron* **24**, 4967 (1968).
16. H.R. Buys, Dissertation, Leiden (1968).
17. C. Altona and A.P.M. van der Veek, *Tetrahedron* **24**, 4377 (1968).
17a. W.J. Adams, H.J. Geise and L.S. Bartell, to be published.
17b. H.J. Geise, W.J. Adams and L.S. Bartell, *Tetrahedron* **25**, 3045 (1969).
18. E.L. Eliel and C.A. Giza, *J. Org. Chem.* **33**, 3754 (1968); R.O. Hutchins, L.D. Kopp and E.L. Eliel, *J. Am. Chem. Soc.*, in press; E.L. Eliel, to be published.
19. E.J. Jacob, H.B. Thompson and L.S. Bartell, *J. Chem. Phys.* **47**, 3736 (1967).
20. K.B. Wiberg, *J. Am. Chem. Soc.* **87**, 1070 (1965).
21. N.L. Allinger, J.A. Hirsch, M.A. Miller, I.J. Tyminski and F.A. Van Catledge, *J. Am. Chem. Soc.* **90**, 1199 (1968); N.L. Allinger, J.A. Hirsch, M.A. Miller and I.J. Tyminski, *J. Am. Chem. Soc.* **90**, 5773 (1968); **91**, 337 (1969).
22. R.H. Boyd, *J. Chem. Phys.* **49**, 2573 (1968).
23. C. Altona, in preparation.
24. C. Altona and H. Hirschmann, *Tetrahedron* **26**, 2173 (1970).
25. D.F. High and J. Kraut, *Acta Cryst.* **21**, 88 (1966).
26. H.J. Geise, Dissertation, Leiden (1964); H.J. Geise and C. Romers, *Acta Cryst.* **20**, 257 (1966).
27. A.D. Cross and P. Crabbé, *J. Am. Chem. Soc.* **86**, 1221 (1964).
28. J. Fishman, *J. Am. Chem. Soc.* **87**, 3455 (1965) and further references cited therein.
29. C. Altona and M. Sundaralingam, *J. Am. Chem. Soc.* **92**, 1995 (1970).
30. G. Dallinga and L.H. Toneman, *Rec. Trav. Chim.* **87**, 795 (1968).
31. A.V. Fratini, K. Britts and I.L. Karle, *J. Phys. Chem.* **71**, 2482 (1967).
32. R.A. Alden, J. Kraut and T.G. Traylor, *J. Am. Chem. Soc.* **90**, 74 (1968).

Nuclear Magnetic Resonance Studies of the Conformations and Conformational Barriers in Cyclic Molecules

F.A.L. Anet
Department of Chemistry
University of California
Los Angeles, California

I want to restrict myself in this chapter to nuclear magnetic resonance (nmr) studies of the conformations of eight-membered carbocyclic molecules. In contrast to the comparatively simple situation in six-membered rings, the conformational possibilities for an eight-membered ring can be very large (*1*). Although nmr is a very powerful tool, the complexity of proton spectra often makes a detailed analysis very difficult. Simplification is generally possible by going to higher spectrometer frequencies and/or by replacing most of the protons by deuterons. Extensive deuterium substitution has the advantage that the relaxation times of the remaining protons are increased and thus sharper lines can be obtained at very low temperatures, i.e., when magnetic-field inhomogeneities no longer contribute dominantly to the line width.

In simple systems, it is often possible to use symmetry properties to exclude rigorously certain conformations. For example, the boat-boat (saddle) conformation (Fig. 1) proposed by Dale (*2*) for cyclooctane can be excluded by nmr measurements on various cyclooctane-d_{14} isotopic isomers (*3*). Also excluded are any conformations, e.g., the twist-boat conformation, which at $-140°C$ have (on the nmr time scale) a time-averaged symmetry of the boat-boat form. The nmr data (*3*) obtained on cyclooctane-d_{14} isotopic isomers show that at $-140°C$ the protons have only two chemical shifts, and that vicinal *trans* protons have the same shifts, whereas vicinal *cis* protons have different shifts.

The data is consistent with the conformation of cyclooctane being the crown form (D_{4d}), *or* any conformation or conformations which are undergoing

15

internal motion rapidly on the nmr time scale at −140°C, so that the time-average symmetry is the same as that of the crown form. Such conformations (4) (Fig. 1) are the chair-chair and twist-chair-chair forms, which can be seen on models (and also according to detailed calculations (4d)) to be very easily converted to the crown form. These forms, however, do not explain the

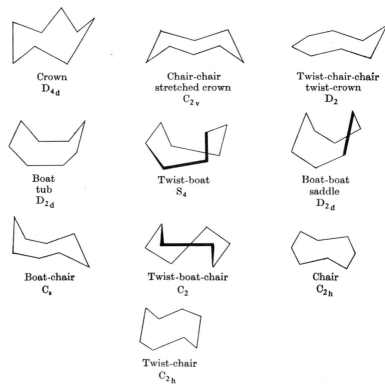

Crown
D$_{4d}$

Chair-chair
stretched crown
C$_{2v}$

Twist-chair-chair
twist-crown
D$_2$

Boat
tub
D$_{2d}$

Twist-boat
S$_4$

Boat-boat
saddle
D$_{2d}$

Boat-chair
C$_s$

Twist-boat-chair
C$_2$

Chair
C$_{2h}$

Twist-chair
C$_{2h}$

Figure 1 Symmetrical conformations for cyclooctane.

behavior of monosubstituted and 1,2-*trans* cyclic substituted cyclooctane derivatives studied by us (5), nor the behavior of 1,1-difluoro- and 1,1,4,4-tetrafluorocyclooctanes studied by Roberts and his co-workers (6).

We came to the conclusion that only the boat-chair form affords a satisfactory explanation for these results (3, 5). At the same time, x-ray work showed that several cyclooctane derivatives indeed had that conformation in the crystal phase (7). A tetrabromocyclooctane, however, has been found to have the chair-chair conformation in the crystal phase (8), so that even though the boat-chair form appears to be of lower energy than the chair-chair form in the simpler derivatives, the difference in energy between these two conformations must be small.

In his early calculations, Hendrickson (*4a*) had found that the boat-chair form was of low energy; in fact, it was the lowest-energy form of all those which he investigated. The ten different types of protons in this form, however, do not seem to be consistent with the previously mentioned nmr work, especially since Hendrickson stated that pseudorotation was not possible in the boat-chair, in contrast to the situation in the previously mentioned chair-chair and boat-boat families. However, as we have shown (*3, 5*), the boat-chair *can* undergo pseudorotation via Hendrickson's C_2 form (conveniently called the twist-boat chair). This pseudorotation looks energetically easy on models, even though the motion required for pseudorotation is not an obvious one when a model is being handled. In his more recent work, Hendrickson (*4d*) has discussed the pseudorotation of the boat-chair. Although his computer program did not allow him to calculate the barrier to pseudorotation, he suggested that the barrier was quite low, in agreement with our view.

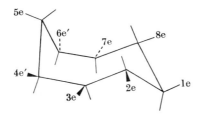

Figure 2   Pseudorotation itinerary in the boat-chair conformation of cyclooctane. Only the equatorial positions are indicated. Set 1: 1e, 2e, 7e, 4a', 5a, 6a', 3e, 8e, 1e, etc. Set 2: 1a, 2a, 7a, 4e', 5e, 6e', 3a, 8a, 1a, etc.

Rapid pseudorotation in odd-membered ring conformations (e.g., cyclopentane and cycloheptane), or in many even-membered ring conformations (e.g., the boat form of cyclohexane, and the boat-boat form of cyclooctane), averages the chemical shifts of all the protons to a single chemical shift in each case. Such is not true in the boat-chair form of cyclooctane, where rapid pseudorotation results in the ten potentially different chemical shifts in the boat-chair being averaged to two chemical shifts (Sets 1 and 2 in Fig. 2). Furthermore, each averaged chemical shift has contributions from equatorial positions. Thus methyl- and *t*-butylcyclooctane show temperature-dependent nmr spectra very similar to cyclooctane itself (*5*). By contrast, methyl- and *t*-butylcyclohexane show essentially temperature-independent spectra (*5*).

The clearest nmr evidence for the boat-chair conformation would be to obtain spectra at temperatures low enough for the rate of pseudorotation to be slow on the nmr time scale, for then there should be ten different chemical shifts in the ratio $1:1:1:1:2:2:2:2:2:2$. So far, it has not been possible to do this (*9*). As stated previously, the barrier to pseudorotation appears to be

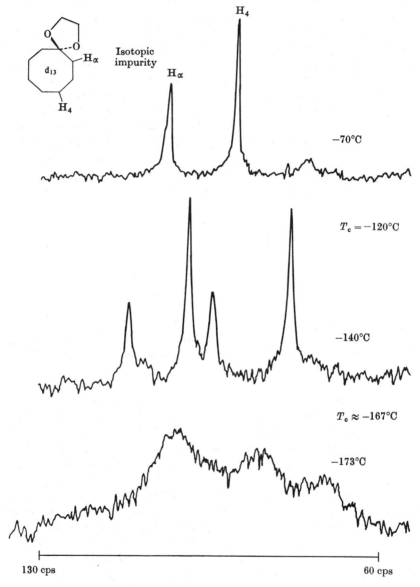

Figure 3 Nuclear magnetic resonance spectra (60 MHz) of a deuterated cyclooctanone ethylene ketal at various temperatures. The scale is cps downfield from TMS.

small, so that even −175°C is not a low enough temperature for this process to be slow on the nmr time scale.

An indirect way to get evidence for pseudorotation is to study substituted cyclooctanes where the barrier may be larger than in cyclooctane itself. A disadvantage of a substituted cyclooctane is that the symmetry of the original molecule is lost. Among the substituents which can be employed, one of the

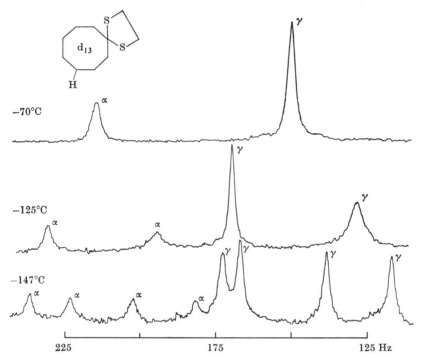

Figure 4   Nuclear magnetic resonance spectra (100 MHz) of a deuterated cyclooctanone ethylene dithioketal at various temperatures.

best is fluorine, which is not only fairly small, but also serves to give a simple spectrum, if the ^{19}F spectrum is obtained. An important advantage is that fluorine chemical shifts are much larger than proton shifts, so that viscosity broadening at very low temperatures is less of a problem than with protons. Roberts and his co-workers have shown that the fluorine-probe method is a very important tool for the study of various conformational problems. With 1,1-difluorocyclooctane, it was shown by Roberts (6) that there are two distinct processes which can be made slow on the nmr time scale on going from room temperature to −175°C. At the lowest temperature, there are four fluorine chemical shifts. This is consistent with the boat-chair conformation if it is assumed that only two kinds of positions are favorable for placing the CF_2 group, the other positions being too sterically strained to be appreciably populated.

Table I Free-Energy Barriers in Cyclooctane and Saturated Derivatives

| | Barriers (kcal/mole) | | |
Compound	Inversion	Pseudorotation	Reference
Cyclooctane	8.1	<5	9, 11
1,1-Difluoro-	7.5	4.9	6
1,1-Ethylenedioxy-	7.6	5.3	9, 10
1,1-Ethylenedithioxy-	8.5	6.6	13
1,1-Dimethyl-	8.0	?	13

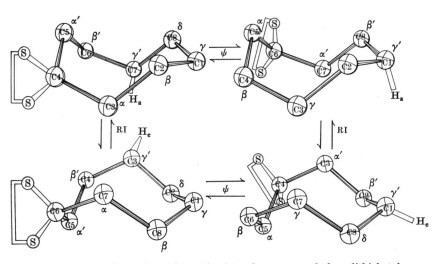

Figure 5   Ring inversion and pseudorotation in cyclooctanone ethylene dithioketal.

We have noted that the ethylene ketal of cyclooctanone also shows two distinct processes (9, 10), but the nmr spectrum at low temperature, even of a deuterated specie (Fig. 3), is too complex to determine how many distinct positions are occupied by the ketal group. The dithioketal of cyclooctanone gave better results (Fig. 4) and showed that there are two processes but only one kind of site for the substituent. With a proton probe in the γ position, it is possible to observe four different but equally populated sites for this probe.

We ascribe the single site for the dithioketal group versus two sites for the CF_2 group to the greater bulk of the dithioketal group compared to the CF_2 group. For the same reason, the barrier to pseudorotation is considerably greater in the dithioketal than in the ketal, where in turn the barrier is greater than in the difluoride (Table I).

The conformational processes in the dithioketal are shown in Fig. 5. The substituent is placed at the 4 (equivalent to 6) position in the boat-chair

Table II   Calculated Dihedral Angles and
            Methyl Strain Energies for Boat-
            Chair Cyclooctane (*4d*)

Dihedral angles (°)		Methyl strain energies (kcal/mole)			
$\omega_{1,2}$	65.0	1e	0.6	3e	0.5
$\omega_{2,3}$	102.2	1a	1.4	3a	7.6
$\omega_{3,4}$	44.7	2e	0.5	4e′	0.5
$\omega_{4,5}$	65.0	2a	5.1	4a′	0.5
				5e	0.5
				5a	8.0

Table III   Barriers to Methyl Rotation

Compound	C⟍C⟋C Angle (°)	Energy barrier (kcal/mole)	Reference
Acetone	116.7	0.80	13
Isobutene	115.9	2.21	14

(Fig. 2), because that is the least hindered position (see energies of methyl substituent for various positions of the boat-chair form, Table II). The processes shown by the vertical arrows in Fig. 5 represent ring inversions (in this process all the dihedral angles change their signs, but not their absolute values). Ring inversion is most probably not a simple process, and probably proceeds via the chair or the twist-chair form followed by fast pseudorotation of the boat-chair, which is initially produced with the substituent in an unfavorable position. Pseudorotation in Fig. 5 and other figures is represented by the letter ψ. Free-energy barriers for cyclooctane and some saturated derivatives are given in Table I. Unlike the barriers to pseudorotation, the barriers to inversion do not vary very much, at least with the substituents shown in the table.

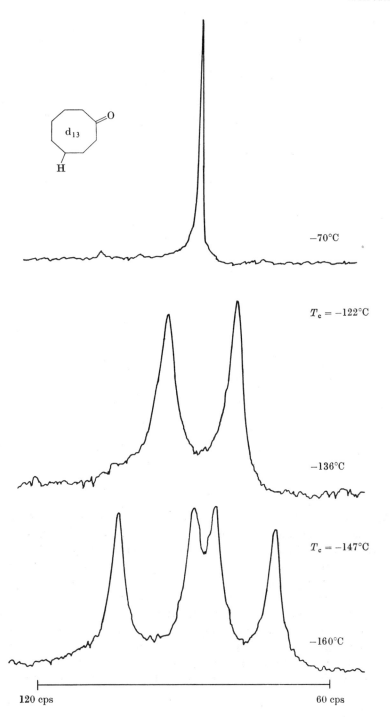

Figure 6 Nuclear magnetic resonance spectra (60 MHz) of a deuterated cyclooctanone at various temperatures. Chemical shifts of γ proton.

In unsaturated derivatives of cyclooctane, the presence of sp^2 hybridized carbon atoms can have three effects: (a) changes in the internal bond angles, (b) changes in nonbonded interactions and (c) changes in eclipsing strain.

For cyclooctanone, changes in the internal bond angles are probably not very important. From the point of view of eclipsing strain, positions 2 and 3 in the boat-chair are best for the placement of the carbonyl group because of the low barrier to internal rotation in ketones (Table III). Nonbonded interactions,

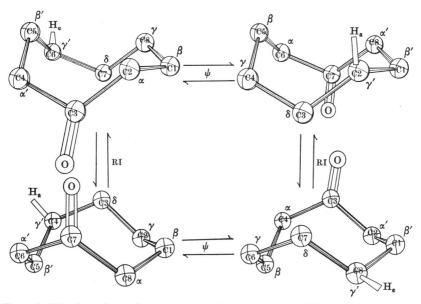

Figure 7 Ring inversion and pseudorotation in cyclooctanone. Effect on γ proton.

however, are best relieved if the carbonyl group is in position 3 or 5 of the boat-chair (see Table II). Thus, these simple considerations* suggest that cyclooctanone should exist in the boat-chair form with the carbonyl group at position 3. The nmr spectra of cyclooctanone-d_{13} with the proton probe in the γ position indeed shows that there are four equally populated sites for the proton (9, 10) (Figs. 6 and 7). With the proton in the δ position, only two lines are obtained at low temperatures and only a single process is seen (12), as expected (Figs. 8 and 9). Of great interest is the very high field position (τ 9.37) of one of the δ protons at low temperatures. We ascribe this to strong shielding from the π system of the carbonyl group at the 3 position on the axial proton at the 7 position of the boat-chair. The barriers to pseudorotation

* Allinger has informed us very recently that computer-based calculations which he has performed indeed indicate that the best chair-boat conformation of cyclooctanone has the carbonyl group in position 3.

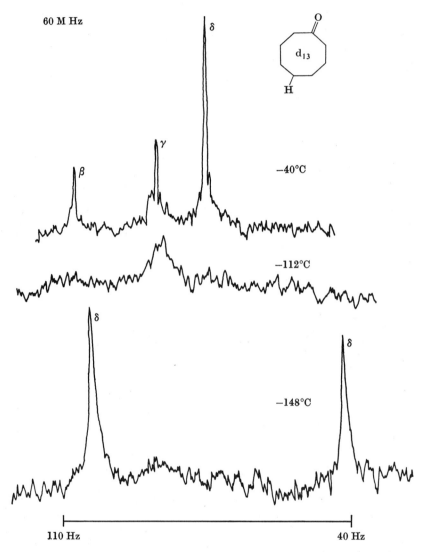

Figure 8   Nuclear magnetic resonance spectra (60 MHz) of a deuterated cyclooctanone.
          Chemical shifts of δ proton.

and ring inversion are fairly close together in cyclooctanone (Table IV). In
5-t-butylcyclooctanone, only pseudorotation can be observed (9). The much
higher barrier (Table IV) than in cyclooctanone can be understood because
the large 5-t-butyl group can make one of the two possible pseudorotation
itineraries very unfavorable. The remaining pseudorotation itinerary, however,

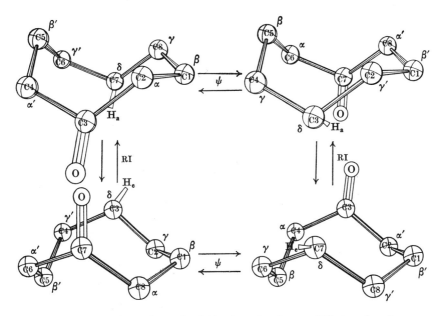

Figure 9 Ring inversion and pseudorotation in cyclooctanone. Effect on δ proton.

although favorable for the *t*-butyl group, may be the lesser favored one for the carbonyl group.

It is perhaps surprising that methylenecyclooctane does not have an analogous conformation to that of cyclooctanone. The low-temperature spectrum shows only one process and only two chemical shifts when the γ proton is observed (*12*) (Fig. 10). Also, the δ proton does not have an abnormal chemical shift. We assign the boat-chair conformation to this compound with the methylene group in the 5 position (Fig. 11). The reason that this conformation is favored is most probably the greater eclipsing strain in olefins as

Table IV Free-Energy Barriers in Unsaturated Cyclooctane Derivatives

Compound	Barriers (kcal/mole)		Reference
	Inversion	Pseudorotation	
Cyclooctanone	7.5	6.3	9, 10
5-*t*-Butylcyclooctanone		8.0	9
Methylenecyclooctane	8.1		12
Cyclooctene	8.2	5.8	9, 10

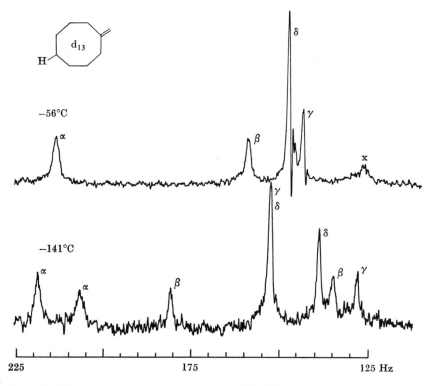

Figure 10   Nuclear magnetic resonance spectra (100 MHz) of a deuterated methylenecyclooctane. Chemical shifts of α, β, γ and δ protons.

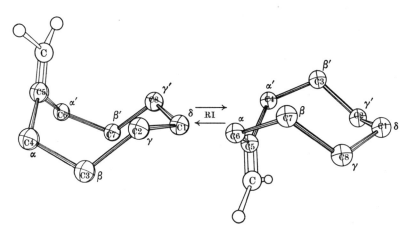

Figure 11   Ring inversion in methylenecyclooctane.

compared to ketones (Table III). The difference is large enough to force cyclooctanone and methylenecyclooctane predominantly (at low temperatures!) into two different conformations, albeit both boat-chairs.

Considerable distortions of the boat-chair have to be made to accommodate the planar system of an internal double bond, as occurs in *cis*-cyclooctene. Energetically, this should not be difficult, because a suitable form, which is of low energy, is one (form h in Fig. 12) which is intermediate between the

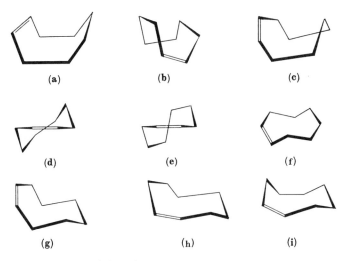

Figure 12 Conformations of *cis*-cyclooctene.

boat-chair and the twist-boat-chair. This conformation possesses no symmetry, in agreement with the fact that both 4 and 5 protons in cyclooctene-d_{13} each have four equally populated sites at very low temperatures (*9*) (Fig. 13). Also, the nmr spectrum of a cyclooctene-d_{12} with protons in the 5 and 6 positions *cis* to one another can be obtained (*9*). It can be deduced (*9*) from the small (1 Hz) coupling constant that the dihedral angle in the system H–C(5)–C(6)–H must be close to 90°. These observations exclude all of the conformations for cyclooctene shown in Fig. 11, with the exception of forms i and h. These two forms are very similar and are very easily interconverted. Both conformations are very close to the conformation of *cis*-cyclooctene predicted by Dale (*15*) on the basis of the known (*16*) (x-ray) structure of a cycloocteneimine.

There are two conformational processes which can be observed in *cis*-cyclooctene (Fig. 13). As usual (Table IV), the lower-energy process is a pseudorotation, but a fuller discussion (*9*) is beyond the scope of this chapter.

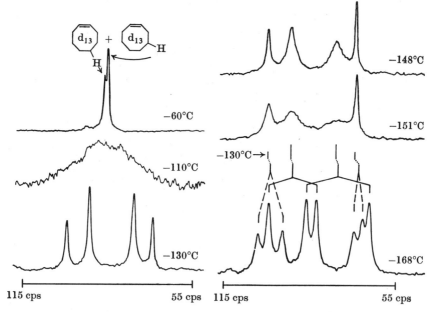

Figure 13   Nuclear magnetic resonance spectra (60 MHz) of a mixture of two different
deuterated cyclooctene species. Chemical shifts of 4 and 5 protons. The major
specie has the proton in the 4 position.

Future applications of nmr to conformational problems will undoubtedly
make increasing use of high-frequency (superconducting solenoid) spectrometers and of ^{13}C magnetic resonance. We hope to apply these tools in the near
future to obtain additional information on some of the problems mentioned
in this lecture, as well as on other conformational problems.

ACKNOWLEDGMENTS

We are grateful to the National Science Foundation for financial support.

REFERENCES

1.   E.L. Eliel, N.L. Allinger, S.L. Angyal and G.A. Morrison, "Conformational Analysis,"
     Wiley (Interscience), New York, 1965; M Hanack, "Conformational Theory,"
     Academic Press, New York, 1965.
2.   J. Dale, I. Laszlo and W. Ruland, *Proc. Chem. Soc.* **1964**, 190.
3.   F.A.L. Anet and M. St. Jacques, *J. Am. Chem. Soc.* **88**, 2585 (1966).
4.   (a) J.B. Hendrickson, *J. Am. Chem. Soc.* **86**, 4854 (1964); (b) K.B. Wiberg, *J. Am.
     Chem. Soc.* **87**, 1070 (1965); (c) M. Bixon and S. Lifson, *Tetrahedron* **23**, 769 (1967);
     (d) J.B. Hendrickson, *J. Am. Chem. Soc.* **89**, 7036, 7043, 7047 (1967).

5. F.A.L. Anet and M. St. Jacques, *J. Am. Chem. Soc.* **88**, 2586 (1966).

6. J.E. Anderson, E.S. Glazer, D.L. Griffith, R. Knorr and J.D. Roberts, *J. Am. Chem. Soc.* **91**, 1386 (1969).

7. P. Groth, *Acta Chem. Scand.* **19**, 1497 (1965); J.D. Dunitz and A. Mugnoli, *Chem. Comm.* **1966**, 166; J.D. Dunitz, as quoted by J.B. Hendrickson, (*4d*); J.V. Egmond and C. Romers, *Tetrahedron* **25**, 2693 (1969).

8. G. Ferguson, D.D. MacNicol, W. Oberhansli, R.A. Raphael and J.A. Zabkiewicz, *Chem. Comm.* 103 (1968).

9. M. St. Jacques, Ph.D. Thesis, University of California, Los Angeles, 1967.

10. F.A.L. Anet, Abstract of papers presented at the Twentieth National Organic Chemistry Symposium of the American Chemical Society, Burlington, Vermont, June 1967.

11. F.A.L. Anet and J.S. Hartman, *J. Am. Chem. Soc.* **85**, 1204 (1963).

12. M. Henrichs and F.A.L. Anet, unpublished work.

13. J.D. Swalen and C.C. Costain, *J. Chem. Phys.* **31**, 1562 (1959).

14. V.W. Laurie, *J. Chem. Phys.* **34**, 1516 (1961).

15. J. Dale, *Angew. Chem.* **78**, 1070 (1966); *Angew. Chem. Int. Ed. Engl.* **5**, 1000 (1966).

16. L.M. Trefonas and R. Majeste, *Tetrahedron* **19**, 929 (1963).

Some Results and Limitations in Conformational Studies of Six-Membered Heterocycles

M.J.O. Anteunis
Department of Organic Chemistry
State University of Ghent
Ghent, Belgium

Since the review by Riddell (*1*) appeared in 1967, a large number of contributions to the conformational analysis of heterocycles have been realized, as is well-illustrated by the papers presented in this book. There are two reasons for the ever-increasing interest in this field. Firstly, the shapes of the molecules envisaged are basically the same as for the more fundamental carbocyclic analogs, and the same principles in both classes are qualitatively governing the phenomena observed. Moreover, the heterocyclic entities are mostly of direct particle interest, as shown by representatives found in nature (e.g., carbohydrates, cyclic peptides, etc.). Secondly, one of the most powerful tools for the study of conformational problems that has become available to all chemists dealing with the subject is nuclear magnetic resonance (nmr), and this technique is particularly helpful for heterocycles, owing to the great simplification in the interpretation of their spectra as compared with the carbocyclic compounds. In the latter case, additional dodges are often needed, of which (per)deuteration (for proton spectra) or (per)fluorination (for ^{19}F spectroscopy) are perhaps the most successful and promising artifices.

Next to the nmr technique, many heterocyclic derivatives may be easily seized chemically, allowing, i.e., interconversion between configurations. In this respect, quantitative gas chromatography must be mentioned as a powerful intervention.

A great deal of our efforts for the moment deals with the conformational behavior of 1,3-dioxanes, 1,3-dithianes and 1,3-thioxanes. In this report, only methods for estimating *equilibria* states between different conformations

X and/or Y = O; S

will be discussed. We have chosen between the compounds mentioned, those which are suitable disubstituted derivatives containing at least one methyl group, such as I and II in such a way that anancomeric situations* are avoided.

$$K_e = \frac{1-x}{x}$$

A	X = Y = O	C	X = O; Y = S
B	X = Y = S	D	X = S; Y = O

I

A	X = Y = O	C	X = O; Y = S
B	X = Y = S	D	X = S; Y = O

II

The conformational energies and thermodynamic parameters derived therefrom are thus always related to that of the methyl group itself. The knowledge of the latter quantity† thus in principle will enable us to extract also the values for the other substituents. This will be especially easy if the energy terms are additive (accumulative) properties, a problem which must be (dis)proven independently.

Investigations with the aid of substituted dioxadecalanes (III) are actually being conducted in our laboratory, and even in those rather complex systems

* The term "anancomeric" was introduced by us (2) as an alternative for "biased," which finds no adequate translation in some other (Germanic) languages. It is derived from the Greek ἀναγκειν: to fix or to provide by some fate or (divine) law.
† In most of the heterocycles actually under discussion these quantities have more or less been determined by other authors, as is the case for 1,3-dioxanes (3, 4) and 1,3-dithianes (5).

the accumulative principle seems to be followed (6), although several interactions must be estimated speculatively. Therefore, predictions of conformational equilibria in III do not differ greatly from experimental settlements.

III

I. EQUILIBRIA DATA IN 1,3-HETEROCYCLIC SYSTEMS USING NUCLEAR MAGNETIC RESONANCE SHIFT DATA

The following discussion only refers to compounds I and II with the exclusion
of 5,5-disubstituted-1,3-dithianes and corresponding 1,3-thioxanes for which
investigations are actually running (7). With respect to the method employed,
Figs. 1a and 2 represent nmr spectral details for two representative compounds.
In Fig. 2, the patterns for H-2 and H-4,6 hydrogens at −86°C are shown for
5-Me-5-i-Bu-1,3-dioxane (100 Mc, carbon disulfide). For both groups of
hydrogens *two* AB systems are found, one system for each of the two possible
conformers of type IIa. These AB systems may be grouped with relative ease,
simply because of the differences in intensities. Thus it is clear that for the
H-4,6 hydrogens, lines 1, 2, 3 and 4 built up one of the AB systems, while
1′, 2′, 3′ and 4′ belong together and form the other system. It was next *assumed**
originally that the bulkiest substituent prefers the equatorial position, thus
the lines 1, 2, 3 and 4 belong to conformation IIA with population x. For the
H-2 hydrogens it is seen that both A parts of the AB systems coincidentally
overlap each other. In most of the *trans*-4,6-dialkyl-1,3-thioxanes, overlapping
is complete, and only one AB system is observed for H-2. We will discuss the
evaluation of equilibria with the aid of Fig. 1. For the H-2 protons of *trans*-
4-Et-6-Me-1,3-dithiane (in CPC†), the two conformers IB are clearly
represented by the two AB systems found at temperatures below the
coalescence temperature (T_c) (Fig. 1a, top).

Above T_c, ring inversion proceeds more rapidly than the spin state inter-
change, and thus only the collapsed spectrum, existing in only a single AB
system, is obtained (Fig. 1a, bottom). At $T < T_c$, we may extract from the
spectrum the values ΔI and ΔII, which represent the shift differences between

* The situation might a priori well be reversed because of the relatively bad knowledge of
synaxial σ–p interactions. This assumption is, however, verified by results obtained by
others (3, 4). Apart from this, we have used other criteria. See below.
† CPC is the code name for the ternary mixture chloroform:pyridine:carbon disulfide,
mostly in 1:1:4 proportion (by volume).

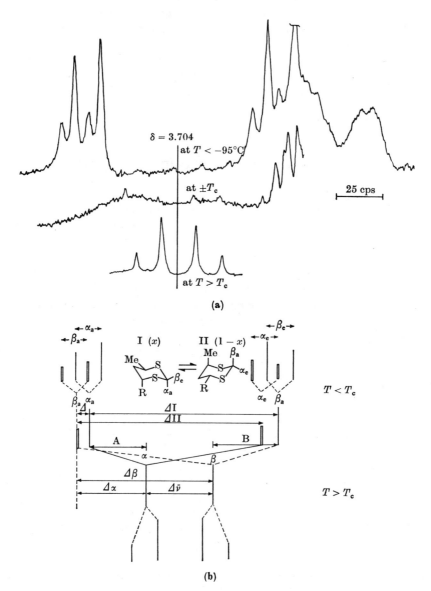

Figure 1 Spectral details for H-2 and H-4,6 in *trans*-4-Et-6-Me-1,3-dithiane (100 Mc in
CPC) at several temperatures, and the theoretical spectra, defining ΔI, ΔII
(at $T < T_c$) and $\Delta\nu$ (at $T > T_c$).

equatorial and axial H-2 hydrogens, in respect of the most and least abundant conformation. These conformations have to be identified before the correct interpretation of the data we will get is possible, and we will discuss this later. At $T > T_c$, we can only measure a weighted mean shift difference between hydrogens α and β represented in Fig. 1b as $\Delta\nu$. When the spatial orientational preference between the two substituents does not differ appreciably, the coalesced pattern (at $T > T_c$) often consists only in a deceptive single line,

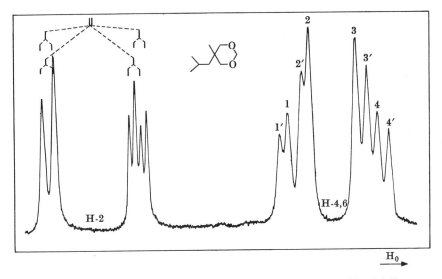

Figure 2 Spectral details for H-2 and H-4,6 at $T < T_c$ for 5-Me-5-i-Bu-1,3-dioxane, showing a "frozen" mixture of two conformations in unequal proportions ($T = -86°C$).

except when one is looking to H-4,6 (geminal) hydrogens (8) as is possible in 5,5-disubstituted derivatives. Vicinal substitutions in this case indeed induce pronounced shift variations (9), so that even for nearly 50:50 equilibria, mostly AB patterns are still found.*

Having the definition of the quantities ΔI, ΔII, $\Delta\nu$, Δ, A, B, $\Delta\alpha$ and $\Delta\beta$ in mind (as follows from inspecting Fig. 1b), it can be shown (11) that equation (1) is valid.

$$K_e = (\Delta I - \Delta\nu)/(\Delta II + \Delta\nu) \tag{1}$$

* These variations due to magnetic anisotropy effects are much smaller—but are real (10)—for more remote substitution. In situations where deceptively simple H-2 patterns are found, the induced shifts by vicinal substituents have the opportunity that an identical study of H-4,6 patterns, as actually under discussion for H-2 hydrogens, can be done. A serious drawback, however, is the decreased accuracy obtained as a result of some suppositions which later on will be introduced. See below.

This follows from the fact that $\Delta\beta = \Delta - \Delta I - B$, $\Delta\alpha = \Delta + A$, and $K_e = (1 - x)/x = $ (intensity line β_e or α_e)/(intensity line β_a or α_a). Here B and A define the positions of lines β and α—the weighted means of the line pairs $\beta_a - \beta_e$ and $\alpha_a - \alpha_e$. We refer to the original paper (11) for the full derivation of equation (1), and for the probable errors introduced by uncertainties on values ΔI, ΔII and $\Delta\nu$.

It is a classical procedure to use the limited values corresponding to those of the individual conformers and which are obtained at some peculiar circumstance, i.e., at $T < T_c$ and to use them together with the value of the mobile conformational mixture present in other circumstances, in order to estimate the composition of that mixture. This has always been done with respect to shift and coupling data. We will show in the next section that this procedure is wrong and that appreciable errors may be introduced. We will show how we have remedied this fallible situation and how we were able to obtain, for the equilibria under study, all thermodynamic parameters with excellent precision. We will also discuss possible sources of errors using the procedure which will be outlined.

A. Unreliability of the Direct Application of Equation (1), and Computation of "True" Equilibrium Constant Values

The values of ΔI, ΔII and $\Delta\nu$ are in most cases so large* that relatively high accuracy was hoped in the calculation of K_e. As the spectra could be taken in a wide temperature range,† it was further expected that from this temperature study not only $\Delta\Delta G°$ values, but also $\Delta\Delta H°$ and $\Delta\Delta S°$ values could have been obtained.

Using the values of ΔI and ΔII (as obtained at $T < T_c$) and that of $\Delta\nu$ at several temperatures (as obtained at $T > T_c$), a bad linear relationship (2) between the computed total free-energy change and temperature was obtained.

$$\Delta\Delta G° = \Delta\Delta H° - T\Delta\Delta S° \qquad (2)$$

Especially in 5,5-disubstituted-1,3-dioxanes and some of the dithianes and thioxanes, some inexplicable and systematic deviations are observed. Tables I and II compile some thermodynamic properties together with correlation coefficients as calculated by the least-squares method and assuming linearity between ΔG and T.

---

* At 100 Mc shift differences (ΔI and ΔII) for H-2 were approximately 50 cps in 5,5-disubstituted-1,3-dioxanes and in *trans*-4,6-disubstituted-1,3-thioxanes; 100 cps in *trans*-4,6-disubstituted-1,3-dithianes, and 50 cps in the corresponding *cis*-4,6-di-substituted-1,3-dithianes.

† The *trans*-4,6-disubstituted-1,3-dioxanes could *not* be studied, because even at −110°C, the conformational equilibrium could not be frozen out (12). For all other compounds, the temperature was extended between about −115°C (20–30°C below T_c) and about +60°C.

Table I   Thermodynamic Values Obtained in the Mobile 4-R-6-Me-1,3-
dithianes and Corresponding Thioxanes[a]

R	ΔI and ΔII constant			ΔI and ΔII exponential		
	$\Delta\Delta H$	$\Delta\Delta S$	Correlation coefficient	$\Delta\Delta H$	$\Delta\Delta S$	Correlation coefficient
			4-R-6-Me-1,3-dithianes			
Et	−367	−0.48	0.9980	−373	−0.52	0.9989
n-Pr	−329	−0.40	0.9940	−354	−0.56	0.9995
i-Pr	−390	+0.06	0.5240	−58	−0.15	0.9680
n-Br	−367	−0.54	0.9910	−369	−0.56	0.9991
i-Br	−223	−0.09	0.8980	−287	−0.54	0.99995
			4,6-RMe-1,3-thioxanes			
Me	1268	0.901	0.992	508	−3.70	0.9995
Et(S)	1590	1.538	0.995	520	−5.01	0.9998
Et(O)	1053	0.001	0.007	518	−2.93	0.9994
n-Bu(S)	1432	1.293	0.998	509	−4.86	0.9997
n-Bu(O)	1036	0.187	0.839	506	−2.84	0.9992
i-Bu(S)	1249	0.871	0.986	508	−3.67	0.9995
i-Bu(O)	1126	0.446	0.926	507	−3.23	0.9994

[a] Using "uncorrected" shift values, and compared with those after correction.

Figure 3 displays graphically some cases encountered in 5,5-disubstituted-
m-dioxanes. In Figs. 4 and 5 some examples of the thioxane and dithiane series
are found. The reason of the inaccuracy is obvious when we look at the
dependence of shift as a function of temperature found in some *anancomeric*
compounds (Fig. 6). The general trend seems, moreover, to be identical
(parallel lines). It can easily be demonstrated (11) that if the observed variations are linear (as suggested by the solid lines in Fig. 6) and applicable on
both ΔI and ΔII values, there is no change at all to be expected on values of K_e.
Closer inspection of Fig. 6, however, reveals a more complex dependence,
because all points lie on a curve which is slightly convex.

The observed dependences may be the result of a great number of causes and
the simplest correlation might well be of exponential nature. This was roughly
checked in a (rather small) temperature range of 20–30°C below T_c and equations (3) and (4) were found to be satisfactory.

$$\Delta I = \Delta I_0 e^{CT} \tag{3}$$

$$\Delta II = \Delta II_0 e^{DT} \tag{4}$$

Table II   Compared Thermodynamic Data in 5,5-Disubstituted-1,3-dioxanes

	$\Delta\Delta G_{300}$	$\Delta\Delta H$	$\Delta\Delta S$	r
	ΔI and ΔII constant			
Me, i-Pr	262	257	−0.02	0.2210
Me, sec-Bu	269	222	−0.17	0.7912
Me, i-Bu	81	74	−0.02	0.3111
Me, $cyclo$-Pe	254	72	−0.61	0.9972
Me, $cyclo$-Hex	261	242	−0.08	0.4632
Me, ϕ	473	109	−1.25	0.9805
	ΔI and ΔII exponentia			
Me, i-Pr	190	364	0.58	0.9985
Me, sec-Bu	178	346	0.56	0.9397
Me, i-Bu	33	132	0.33	0.9984
Me, $cyclo$-Pe	117	270	0.51	0.9999
Me, $cyclo$-Hex	130	362	0.57	0.9995
Me, ϕ	316	396	0.27	0.9908

The temperature range, which is experimentally obtainable, is much too small to allow accurate evaluations of ΔI_0, ΔII_0, C and D. This is not the case in the range where $\Delta\nu$ ($T \geqslant T_c$ up to 60–70°C) may be measured, and therefore the following assumptions and calculations were made. By analogy we may write

$$\Delta\nu = \Delta\nu_0 e^{BT} \tag{5}$$

and from a $\ln\Delta\nu$ versus T plot, values of $\Delta\nu_0$ and B may be obtained with reasonable accuracy (least-squares method gives correlation coefficients always greater than 0.999). Because at $T = 0$ we have $K_e = 0$, it follows from equation (1) that the so-obtained $\Delta\nu_0 = \Delta I_0$, and therefore equation (3) may now be rechecked after calculation of $\Delta\nu_0$. The constancy of C was found to be excellent in all cases.

Now a second and not directly controllable assumption must be introduced—that in the temperature range actually under consideration, the differences of the shifts $\Delta I - \Delta II$ do not change. This was always verified in the 20–30°C temperature range below T_c, and moreover this seems logical because the lines (as shown in Fig. 6) are parallel.*

* Parallelism was also checked for other compound classes. In 1,3-dithianes only cis-4,6-disubstituted derivatives could be prepared and we tacitly accept the behavior to be universal (e.g., also applicable in $trans$-substitution etc.).

That the constancy of $\Delta\mathrm{I} - \Delta\mathrm{II}$ is only valuable in a rather restricted temperature range is obvious because we have

$$\ln K_\mathrm{e} = -(\Delta\Delta H°/RT) + (\Delta\Delta S°/R)$$

and thus $\lim K_\mathrm{e}(T \to \infty) \simeq 1$ if $\Delta\Delta S°$ is very small (which turns out to be the case). Therefore at $T = \infty$ it follows from equation (1) that $\Delta\nu = \frac{1}{2}(\Delta\mathrm{I} - \Delta\mathrm{II})$.

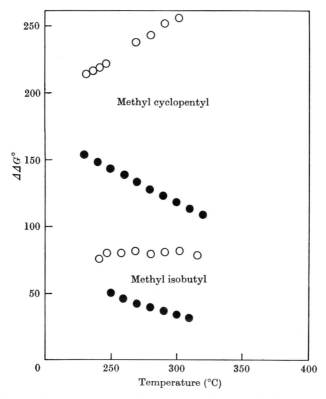

Figure 3   Failure ("without correction" o o o) and applicability ("with correction" ● ● ●) of the expected linear relationship of $\Delta\Delta G$ and T in 5,5-disubstituted-1,3-dioxanes.

If $\Delta\mathrm{I} - \Delta\mathrm{II}$ is constant, this means that $\Delta\nu$ must be definite, which is in contradiction with expression (5). Thus clearly, the latest assumption is not valid at extreme temperature variations.

We are now able to calculate "real" values of $\Delta\mathrm{I}$ and $\Delta\mathrm{II}$ at any temperature (above T_c) and also $\Delta\nu$ values (below T_c). Direct application of equation (1) gives us the "true" K_e values, and from this, new $\Delta\Delta G°$, $\Delta\Delta H°$ and $\Delta\Delta S°$ values (Tables I and II). An excellent linear regression between $\Delta\Delta G°$ and T

is now obtained, as illustrated in Figs. 3–5 and 8, except for some rare cases (see Section B). We searched for an independent proof for the correctness of our procedure. In 5-Me-5-ϕ-1,3-dioxane, the two methyl signals at $T < T_c$ are separated enough to allow the estimation of K_e from weighing of the surfaces. This gives us (at $T = -90°C$) the value of $K_e = 0.410$. Calculation with the aid

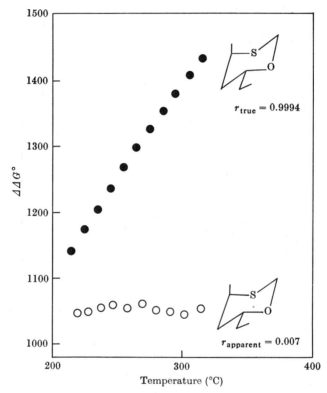

Figure 4   Apparent and "true" $\Delta\Delta G$ versus T relationship in some 4,6-*trans*-disubstituted-1,3-thioxanes.

of data of Table II gives us, for the "corrected" K_e value at $-90°C$, $K_e = 0.411$, which is more than excellent agreement. The K_e from uncorrected values of $\Delta\Delta H°$ and $\Delta\Delta S°$ disagrees with experiment.

B. Exceptions

No satisfactory results were obtained for at least two cases. One of these is found in 5-Me-5-*t*-Bu-1,3-dioxane (Fig. 9), for which an unreasonable $\Delta\Delta G°$ value is found (3.4 kcal/mole; expected (*3, 4*): ~1.0 kcal/mole). Equilibrium in this case, however, is expected (4) to be shifted rather greatly towards the conformation having the t-Bu group in equatorial position. As a result, it is a priori not impossible to be unable to detect a true coalescence temperature, as indeed is the case (up to $-100°C$). Because for all others in this series T_c is about $-60°C/-70°C$ (at 100 Mc), we would accept that at $-100°C$ the conformations are frozen out. In reality, however, we may expect two transition

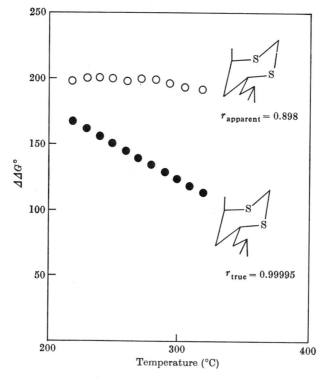

Figure 5   Apparent and "true" $\Delta\Delta G$ versus T relationship in some 4,6-*trans*-disubstituted-1,3-dithianes.

temperatures for those cases where conformational population is so widely different, resulting in two rather different barriers for inversion, depending on what path is followed (from one conformer to the other, or vice versa). Therefore the nonreliability of our calculations in 5-Me-5-t-Bu-1,3-dioxane can be the result of trivial causes. The situation, however, is entirely different in the anomalous case represented in 4-i-Pr-6-Me-1,3-dithiane $(\Delta\Delta G° = -12.3(?)$ cal/mole$)$. We have good reasons to believe that the main cause is the presence of appreciable amounts of flexible forms even at room temperature. This follows from the findings given here:

(1) The geminal coupling constant ($^2J_2$) value decreases 1 cps (in absolute value) with a 100°C temperature rise. This behavior is abnormal, the other members of the 1,3-dithianes not acting in that way. If the behavior of $^2J_2$

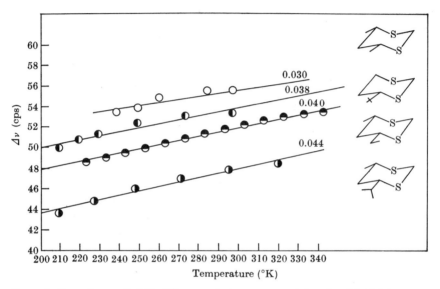

Figure 6 Dependence of shift differences between equatorial and axial H-2 in some anancomeric 1,3-dithianes. (Numbers in figure indicate cps/°K.)

is identical with that observed in 1,3-dioxanes, this means that increasing amounts of flexible forms occur with temperature elevation. It is indeed clear (*13*) from Fig. 10 and from the inspection of Dreiding models that the normal $^2J_2$ value of −6.0 cps drops (in absolute value) to about −2.8 cps for a twist form (see 4-adamantyl-6-*t*-butyl-1,3-dioxane) with the axis of (pseudo)-symmetry passing through C-2. Even classical boot forms have appreciably different $^2J_2$ values (4.5–4.7 cps). We are working out these data on a quantitative scale, for which we need some more "true" flexible model compounds, but it is clear that in the measurement of $^2J_2$ values, a new criterion can be found for the detection or even quantization of flexible forms.*

* Note added in proof: Very recently we have used the limiting values for $^2J_2$ in 1,3-dioxanes in order to calculate twist population in some *trans*-4,6-diR-1,3-dioxanes (M. Anteunis and G. Swaelens, *Organ. Magn. Res.*, in press).

(2) Lambert has found (14) that in equilibria of equi-energetic conformers the ratio J_{trans}/J_{cis} can be used for the detection of the presence of flexible forms. In all *trans*-4,6-diR-1,3-dithianes, this ratio is indeed $\geqslant 3$. In the *i*-Pr derivative, however, the ratio is only 2.3.

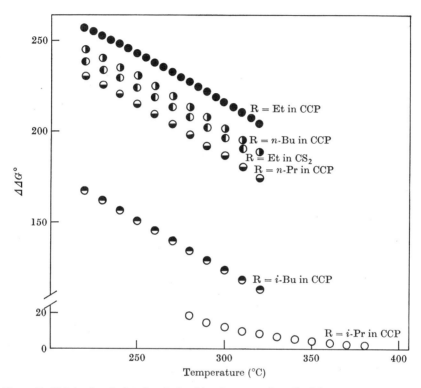

Figure 7 Obtained calculated relationships between "true" $\Delta\Delta G$ values and T in 4-R-6-Me-1,3-dithianes.

C. Discussion of Precision of the Method and Determination of Signs of the Thermodynamic Parameters

As seen, it is possible to use, in some cases, not only data extracted from H-2 patterns, but also those from H-4,6 patterns. In some cases, i.e., for equilibria close to 50:50 (as in 5,5-dialkyl-*m*-dioxanes) only the latter patterns can be used. Due to the vicinal branching, the shifts are very sensitive to small changes in substitution, being reflected in the values $\Delta I - \Delta II$.

For H-2 hydrogens in thioxanes, the fact that $\Delta I - \Delta II \simeq 0$ is in favor of the precision of the resulting thermodynamic calculated quantities, and this is still good in dithianes where $\Delta I - \Delta II \simeq 4$ cps. The worst cases are the dioxanes (5,5-disubstitution) where $\Delta I - \Delta II \simeq 10$ cps (for H-4,6) and this invalidates

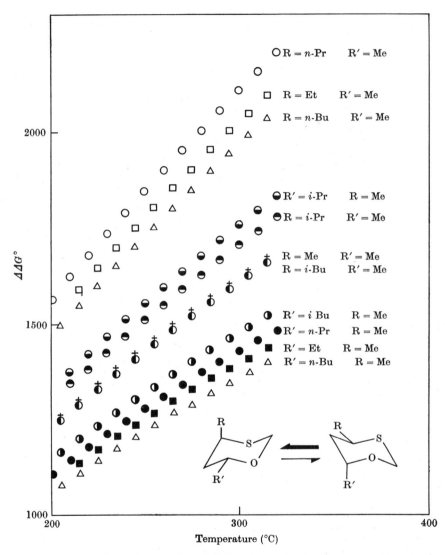

Figure 8   Linear relation between "true" $\Delta\Delta G$ values and T in 1,3-thioxane derivatives.

the accuracy of the results. It is always found that the results from H-2 hydrogens are more precise than from H-4,6 hydrogens.*

* The analyses are always done in accepting pure AB systems. An additional error may be introduced because the hydrogens under consideration are involved in long-range coupling phenomena (15–18). This is even more pronounced for H-4,6 than for H-2 and in 5,5-diR-m-dioxanes, we have in fact an AA'BB'XX' system in which not only J_{AX} and $J_{A'X}$, but also J_{AB} or $J_{AB'}$ and even (18) $J_{AA'}$ may be real.

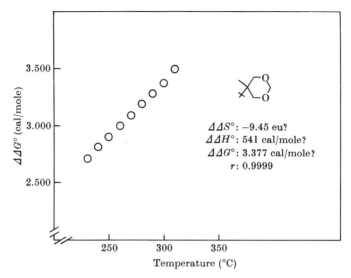

$\Delta\Delta S°$: −9.45 eu?
$\Delta\Delta H°$: 541 cal/mole?
$\Delta\Delta G°$: 3.377 cal/mole?
r: 0.9999

Figure 9   Failure of predicting correct $\Delta\Delta G°$ value in 5-Me-5-t-Bu-1,3-dioxane using H-2 hydrogens.

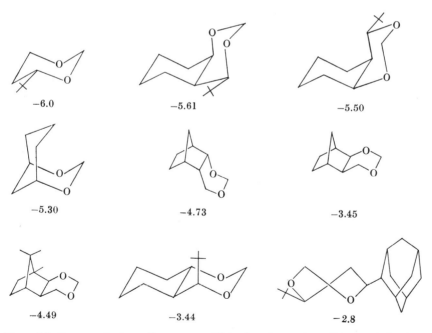

−6.0   −5.61   −5.50

−5.30   −4.73   −3.45

−4.49   −3.44   − 2.8

Figure 10   Some geminal coupling constant ($^2J_2$) values (in cps) in 1,3-dioxanes belonging to different families.

The next problem we will tackle is the ambiguity of the calculations of equilibrium constants which can subsist as the result of the ignorance of what AB systems must be attributed to what conformation. Even more troublesome is the observation that the location of the signal corresponding to the equatorial hydrogen is not always found to be systematically the same in sequence with respect to the location of the signal due to the axial counterpart. This directly influences the relative *signs* of ΔI and ΔII, and of course this results in completely different values for K_e when applying equation (1). As a general procedure we always calculate the K_e sets *for all possible combinations.*

If two AB systems are visible, this results in sixteen possible combinations when the A parts cannot be definitively assigned to the respective B parts. This is, however, possible only in the case that the intensities of the two systems are not differentiated enough. Otherwise, the total of combinations remounts to eight combinations only.

Of all the sets, usually a great number can be eliminated simply on the basis of the bad correlation coefficients obtained. If the correct combinations of A towards B parts are picked out, the question remains whether one of the AB parts belong to one conformation or to another. That is, what conformation occurs in the lowest population?*

Some a priori guesses, of course, are possible, but it is safe to check these assumptions whenever possible. Several criteria may be followed, some of which can also decide upon the correct sign of ΔI and ΔII to be taken and of which the following are the most important:

(1) The shifts of the methyl signals can be predicted, as follows from separated studies on model compounds, and one can easily attribute equatorial and axial substitution when both spectra are confronted. Here shifts of the Me groups follow the traces found for the ring protons (see (5) below).

(2) An axial methyl group is broadened when vicinal axial ring proton(s) (see (1) above) are present. This is due to long-range coupling following a planar zigzag path (*19*).

(3) An axial methyl group is involved in an apparently larger coupling (0.1–0.2 cps) with the vicinal ring hydrogen than is an equatorial methyl group. This was verified (*22*) for Me-2 and Me-4 (but seems not to be followed for Me-5 substitution). This seems to be held also in cyclohexane derivatives (*20*) and until now the reason is not well-understood. Although the differentiation might well be only apparent and be the result of deviation from pure first-order character, identical coupling values were found for Me-5e and Me-5a in thioxanes and dithianes, where the same argument should hold.

* Changing the assignation of conformers will result in a change in the signs of $\Delta\Delta H^\circ$ and $\Delta\Delta S^\circ$.

(4) Typical long-range couplings involving ring protons can answer the question of its spatial orientation. Due to the greatly diminished resolution of low-temperature spectra, however, this criterion is not always very helpful.

(5) We have a fairly good idea now, from separate studies on model compounds, where equational ring hydrogens have to be expected in relation to their axial counterparts. If H-e absorbs at higher field values, it follows from Fig. 1 that ΔI (or ΔII) is to be taken as positive.

This is, for H-2 hydrogens, the most frequently encountered situation (2, 8, 11, 21) (but is reversed to what is "currently" observed for other positions and for carbocyclic derivatives). This is, however, not systematically true and especially voluminous synaxial interactions may reverse the situation (15).

For H-4,6 hydrogens—influenced by vicinal 5-substitution in the compounds under study—the order is even more capricious. High-branched equatorial C-5 substitution may dramatically change the sequence (which now "normally" is (2) that δ(H-4e) > δ(H-4a)), as is, e.g., the case (9) in 5-Me-5-t-Bu-1,3-dioxane.

Other even more unpredictable cases are obviously phenyl-substituted heterocycles, because small orientation (rotamer) differences cause great shift variations of the ring hydrogens (13). All these more or less extra influences through magnetic susceptibility are the reasons why it is extremely dangerous to try to use *absolute shift* parameters (even with respect to internal standard added) and to compare them with absolute shifts obtained from model compounds, in the hope to obtain some insight on conformational equilibria. An extensive study (12, 13) on m-dioxanes, of this problem has shown that it is absolutely meaningless to try to do so, because too many unknown parameters are involved, and the agreement with the real situation of the results, neglecting subtle influences, is only fortuitous.

II. THE FAILURE OF COUPLING CONSTANT CRITERIA FOR THE EVALUATION OF CONFORMATIONAL EQUILIBRIA

Besides shift criteria, the interpolation of apparent coupling constants between limited "model" values is a second widely used procedure for the evaluation of conformational equilibria. Table III, obtained for 4-ϕ-6-alkyl-1,3-dioxanes (13), immediately illustrates the nonreliability of this method. The values of population fractions, x, obtained from interpolation of apparent vicinal coupling values involving essential H-5a and H-5e hydrogens with H-4 and H-6, give widely diverging results. Moreover, some different "model" compounds (Table III, left side) have dramatically different limit values.

The nonreliability is perhaps best illustrated when comparing the 4-ϕ and 4-p-nitrophenyl derivatives. Due to the more pronounced dipole leveling in the axial 4-p-nitrophenyl derivative when compared with the parent unsubstituted phenyl derivative, one should expect a smaller x value for the former.

Table III   Calculated and Compared Fractions of Population x Obtained with the Aid of Vicinal Coupling Constant Values in Some 4-Aryl-6-alkyl-1,3-dioxanes

Equilibria from observed J values:

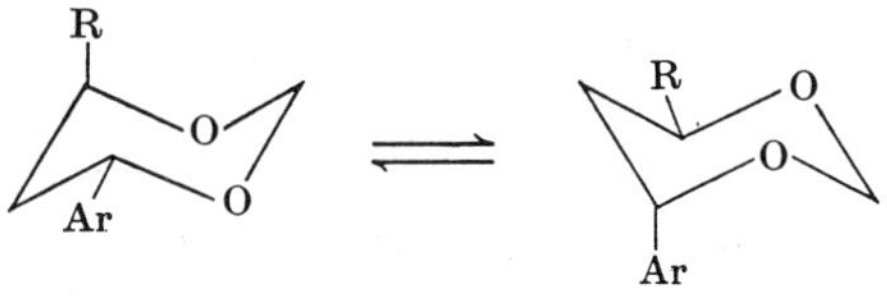

$x \quad\rightleftharpoons\quad 1-x$

R	Ar	Solvent	$J(5a,4)$	$J(5a,6)$	$J(5e,4)$	$J(5e,6)$
Me	ϕ	CS_2	0.48	0.44	0.48	0.55
Me	p-$NO_2\phi$	$CDCl_3$	0.62	0.70	0.38	0.62
		CCl_4	0.71	0.74	0.85	0.66
Et	ϕ	CS_2	0.30	0.55	0.42	0.45
n-Pr	ϕ	CS_2	0.42	0.48	0.59	0.55
neo-Pe	ϕ		0.50	0.56		0.41
i-Bu	ϕ		0.44	0.46	0.43	0.42
i-Pr	ϕ			0.34	0.24	0.22
CF_3	ϕ	CS_2	0.99	0.92	1.19	0.68
		Pyridine	0.62	0.89	0.39	0.64
Me	$CH_2\phi$	CS_2	0.51	0.65	0.54	0.61

Models

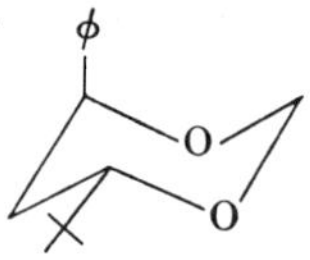

ϕ:
$J(5a,6a) = 11.62$
$J(5e,4e) = 1.95$
$J(5e,6a) = 2.67$
$J(5a,4e) = 5.92$

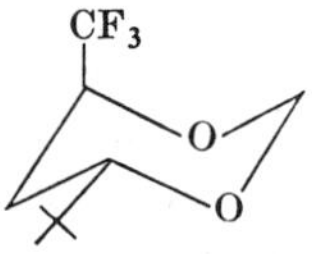

CF_3:
$J(5a,4a) = 12.02$
$J(5e,6e) = 1.50$
$J(5e,4e) = 2.79$
$J(5a,6e) = 7.36$

Hence, the mean result from coupling values is the contrary. Moreover, x "seems" to be higher in the more apolar carbon tetrachloride than in chloroform. This is not possible.

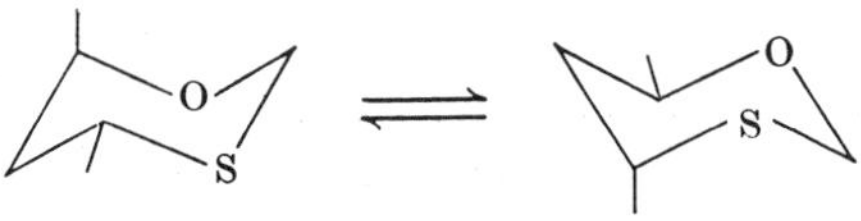

Figure 11   Equilibrium from coupling constant values, calculated in $trans$-4,6-dimethyl-1,3-thioxane ($\Delta\Delta G° = 1.10$ kcal/mole, $K_e = 85/15$, calculated value = $2.9 - 1.7 = 1.2$).

Another example is striking. From apparent coupling values in the mobile system, $trans$-4,6-diMe-1,3-thioxane, and using appropriate reference values, the equilibrium (Fig. 11) was calculated to be 85% in favor of the methyl group being axial and next to the sulfur side. The corresponding $\Delta\Delta G°$ value of 1.10 kcal/mole seemed to be in excellent accordance with independently obtained conformational energies for the Me-4 group in m-dioxane (2.9

kcal/mole) (*4*) and *m*-dithiane (1.7 kcal/mole) (*5*). However, the more reliable shift criterion has meanwhile shown (*7*) that $\Delta\Delta G°$ is larger, being in fact 1.62 kcal/mole.

We may conclude that at the moment conformational evaluations using coupling constant data can only be taken as crude approximation, notwithstanding the increasing effort put into a reevaluation of, e.g., the Karplus rule (*21*). We may fear that this dubious situation is far from settled; the only thing which for the moment can be said with respect to most of the criteria discussed in this paper unfortunately sounds like, "each compound behaves as an individual, impeding often sophisticated combinations in order to obtain precise conformational information."

ACKNOWLEDGMENTS

I wish to express my thanks to all of my collaborators, who did most of the experimental work presented in this paper—Drs. G. Swaelens, J. Gelan, E. Coene and D. Tavernier, and Mr. F. Borremans and Mr. G. Verhegge.

REFERENCES

1. F.G. Riddell, *Quart. Revs.* **21** (3), 364 (1967).
2. M. Anteunis, D. Tavernier and F. Borremans, *Bull. Soc. Chim. Belges* **75**, 396 (1966).
3. F.G. Riddell and M.J.T. Robinson, *Tetrahedron* **23**, 3417 (1966).
4. E.L. Eliel and M.C. Knoeber, *J. Am. Chem. Soc.* **88**, 5347 (1966).
5. E.L. Eliel and R.O. Hutchins, *J. Am. Chem. Soc.* **91**, 2703 (1969).
6. G. Swaelens and M. Anteunis, *Bull. Soc. Chim. Belges* **78**, 321 (1969); *Tetrahedron Letters* **1970**, 561.
7. J. Gelan and M. Anteunis, unpublished results.
8. E. Coene and M. Anteunis, *Bull. Soc. Chim. Belges* **79**, 37 (1970); *Tetrahedron Letters* **1970**, 595.
9. M. Anteunis, E. Coene and D. Tavernier, *Tetrahedron Letters* **1966**, 4579.
10. D. Tavernier and M. Anteunis, *Tetrahedron Letters* **1966**, 5851; and unpublished results together with G. Swaelens.
11. J. Gelan and M. Anteunis, *Bull. Soc. Chim. Belges* **78**, 599 (1969); see also **79**, 313 (1970).
12. D. Tavernier and M. Anteunis, unpublished results.
13. G. Swaelens and M. Anteunis, unpublished results.
14. J.B. Lambert, *J. Am. Chem. Soc.* **89**, 1836 (1967); *Tetrahedron Letters* **1967**, 4755. H.R. Buys (*Rec. Trav. Chim. Pays-Bas*, **88**, 1003 (1969)) has shown that the "*R*-value" can be related to the torsional ring angle τ by the expression $R = J_{trans}/J_{cis} = (3-2\cos^2\tau)/4\cos^2\tau$. The value for τ found in the *trans*-dithianes are generally 62°, except in the methylisopropyl derivative, where it is 59°. For a recent discussion of these torsional angles in 1,3-heterocyclic derivatives on the basis of a modified Lambert criterion see J. Gelan, G. Swaelens and M. Anteunis, *Bull. Soc. Chim. Belges* **79**, 321 (1970). See also N. de Wolf and H.R. Buys, *Tetrahedron Letters* **1970**, 551.
15. D. Tavernier and M. Anteunis, *Bull. Soc. Chim. Belges* **76**, 157 (1967).

16.   J. Feeney, M. Anteunis and G. Swaelens, *Bull. Soc. Chim. Belges* **77**, 121 (1968).

17.   J. Gelan and M. Anteunis, *Bull. Soc. Chim. Belges* **77**, 447 (1968).

18.   M. Anteunis and A. De Bruyn, unpublished results.

19.   M. Anteunis and D. Tavernier, *Bull. Soc. Chim. Belges* **76**, 432 (1967).

20.   A. Segre and J.I. Musher, *J. Am. Chem. Soc.* **89** (3), 706 (1967).

21.   J. Gelan and M. Anteunis, *Bull. Soc. Chim. Belges* **77**, 423 (1968); see also M. Anteunis, *Bull. Soc. Chim. Belges* **75**, 413 (1966).

22.   M. Anteunis, *Bull. Soc. Chim. Belges* **79** (1970), in press.

Die konformative Beweglichkeit der gesättigten Siebenringverbindungen. Ringinversion, Version und Pseudorotation des 5,5-Dimethyl-3,3,7,7-tetradeuterium-1,2-dithiacycloheptans

K. von Bredow,* H. Friebolin† und S. Kabuss

Institut für Elektrowerkstoffe der Fraunhofer-Gesellschaft Freiburg und dem Chemisches Laboratorium der Universität Freiburg, Deutschland

Für die Cycloheptanmolekel gibt es 4 Konformationen, die sich durch ihre Symmetrieeigenschaften auszeichnen: die Sessel(S)-, Twistsessel(TS)-, Twistwannen(TW)- und Wannen(W)-Konformationen. Die Sessel- und Wannen-Konformationen haben die Symmetrie C_{1v}; die Twistsessel- und Twistwannen-Konformationen haben die Symmetrie C_2. Für die Sym-

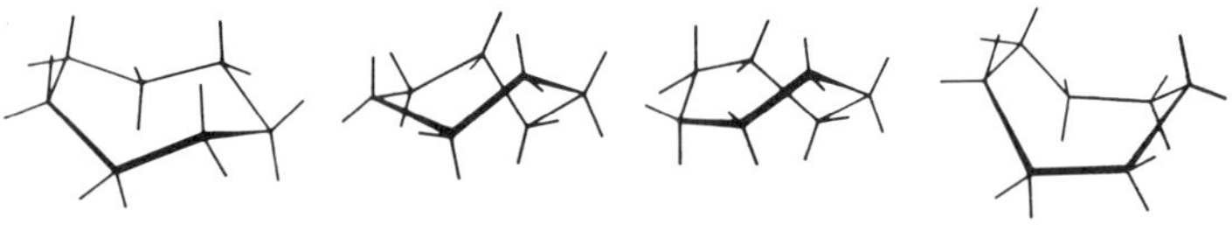

Abbildung 1 Die vier durch ihre Symmetrie ausgezeichneten Konformationen des Cycloheptans.

metrieelemente gibt es jeweils sieben verschiedene Anordnungsmöglichkeiten, da jedes beliebige der 7 C-Atome auf dem Symmetrieelement angeordnet werden kann.

Ihre neue Adressen:
* Geigy AG, Basel/Schweiz.
† Institut für Makromolekulare Chemie der Universität Freiburg.

51

Für jede Lage des Symmetrieelements gibt es zudem zwei *inverse* Konformationen, die sich durch die Vorzeichen sämtlicher Torsionswinkel unterscheiden. Beim Cycloheptan sind somit je 14 Konformationen mit S-, TS-, W- und TW-Konformation zu berücksichtigen. Die Unterscheidung der einzelnen Konformationen erfolgt durch zwei Indices. Der erste dieser Indices, eine Ziffer, gibt an, durch welches C-Atom das Symmetrieelement verläuft; der zweite Index, ein Plus- oder Minus-zeichen, zeigt an, um welche der beiden inversen Konformationen es sich handelt (*1*). S_{5+} bedeutet somit dass das Molekül Sesselkonformation hat, die Spiegelebene durch C-Atom 5 geht, und der Torsionswinkel ω_5 (Torsionswinkel zwischen C-5 und C-6) positiv ist (*1*).

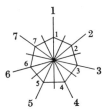

Abbildung 2 Die sieben verschiedenen Anordnungsmöglichkeiten des Symmetrieelements.

Sind, wie beim unsubstituierten Cycloheptan, alle Liganden gleich, ist jeweils der Energieinhalt aller 14 Konformationen mit derselben Form, also z.B. aller S- (bzw. TS-, TW-oder W-) Konformationen gleich. Bei substituierten Cycloheptanen können im ungünstigsten Falle alle Konformationen unterschiedlichen Energieinhalt haben, jedoch sind bei allen Molekülen mit gleichen geminalen Liganden jeweils die beiden inversen Konformationen energiegleich.

Hendrickson (*2*), sowie Bixon und Lifson (*3*), haben mit Modellrechnungen für Cycloheptan den Energieinhalt der Sessel-, Twistsessel-, Twistwannen- und Wannen-Konformationen berechnet. Energieärmste Konformation ist sowohl nach Hendrickson als auch nach Bixon und Lifson die Twistsessel-Konformation; die Konformationsenergien von Sessel-, Twistsessel- und Wannenkonformation sind in der folgenden Tabelle angegeben.

	Twistsessel (kcal/mole)	Sessel (kcal/mole)	Twistwanne (kcal/mole)	Wanne (kcal/mole)
Hendrickson	0	2,16	2,49	3,02
Bixon und Lifson	0	0,67	2,64	2,40

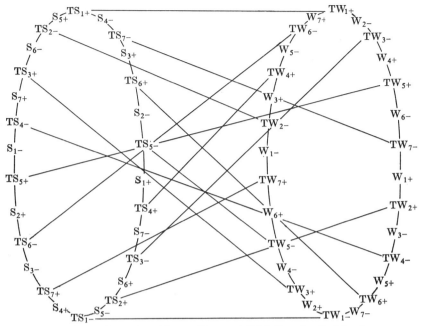

Abbildung 3 Die konformativen Umwandlungsmöglichkeiten des Cycloheptans.

Die folgenden Möglichkeiten der gegenseitigen Umwandlung der symmetrie-ausgezeichneten Konformationen des gesättigten Siebenringes sind zu unterscheiden:

(1) Die Umwandlungen der Sessel (S)- und Twistsessel (TS)-Konformationen ineinander, die *Pseudorotation* der Sesselfamilie: S $\rightleftharpoons$ TS. Übergangskonformation ist nach Hendrickson beim Cycloheptan der Sessel, $E = 2{,}16$ kcal/mole.

(2) Die Umwandlungen der Wannen (W)- und Twistwannen (TW)-Konformationen ineinander, die *Pseudorotation* der Wannenfamilie: W $\rightleftharpoons$ TW. Übergangskonformation ist nach Hendrickson die Wanne, $E = 0{,}53$ kcal/mole.

(3) Die Umwandlungen von Konformationen der Sesselfamilie in solche der Wannenfamilie die *Versionen*: S $\rightleftharpoons$ W, S $\rightleftharpoons$ TW, TS $\rightleftharpoons$ W, TS $\rightleftharpoons$ TW. Als günstigste Übergänge berechnete Hendrickson für Cycloheptan die Übergänge $TS_{i+} \rightleftharpoons TW_{i+}$ zu 8,5 kcal/mole. Die Konformation der Übergangsformen ist halbsesselähnlich.

(4) Die gegenseitigen Umwandlungen der inversen Konformationen, die *Ringinversionen*: $S_{i+} \rightleftharpoons S_{i-}$, $TS_{i+} \rightleftharpoons TS_{i-}$, $W_{i+} \rightleftharpoons W_{i-}$, $TW_{i+} \rightleftharpoons TW_{i-}$. Eine Ringinversion kann auf zwei verschiedenen Wegen erfolgen: (a) durch Pseudorotation innerhalb einer Familie; und (b) durch eine Folge von zwei Versionsübergängen mit zwischengelagerter Pseudorotation innerhalb der anderen Familie, z.B., $TS_{1+} \rightleftharpoons TW_{1+} \rightleftharpoons W_{2-} \cdots W_{7+} \rightleftharpoons TW_{1-} \rightleftharpoons TS_{1-}$.

Bei zwei gleichsinnig erfolgenden Pseudorotationsschritten, sowohl in der Sesselfamilie als auch in der Wannenfamilie, vertauschen entweder zwei geminale Liganden formal ihre Plätze oder *ein* Torsionswinkel ändert sein Vorzeichen: (a) bei den Pseudorotationsschritten $S_{1+} \rightleftharpoons TS_{5-} \rightleftharpoons S_{2-}$ (vergleiche Abb. 4) vertauschen H_{5a} und H_{5e} formal die Plätze; und (b) bei den Übergängen $W_{1-} \rightleftharpoons TW_{2-} \rightleftharpoons W_{3+}$ sind dies H_{2a} und H_{2e}. Im Falle der Pseudorotation von $TS_{1+} \rightleftharpoons S_{5+} \rightleftharpoons TS_{2-}$ ändert ω_1 sein Verzeichen, bei $TW_{1+} \rightleftharpoons W_{2-} \rightleftharpoons TW_{3-}$ ist dies ω_5. Nur bei den Ringinversionen werden *alle* axialen und äquatorialen Liganden formal ausgetauscht, d.h. im PR-Spektrum äquivalent.

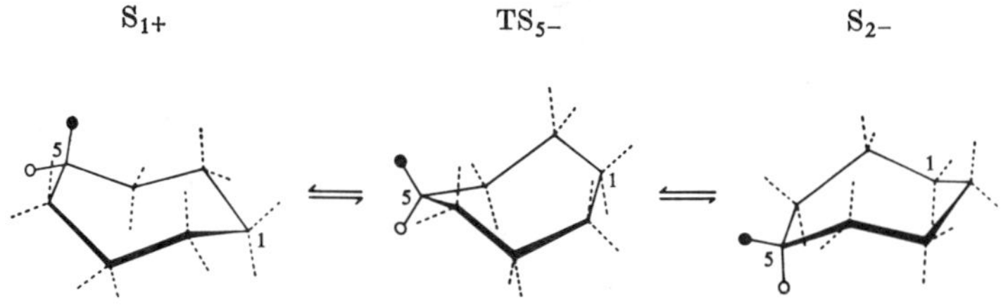

$$S_{1+} \qquad\qquad TS_{5-} \qquad\qquad S_{2-}$$

Abbildung 4 Darstellung der Pseudorotationsschritte $S_{1+} \rightleftharpoons TS_{5-} \rightleftharpoons S_{2-}$; H_{5a} und H_{5e} vertauschen formal ihre Plätze.

Da die Höhe der Pseudorotationsbarriere sehr niedrig ist (vergleiche Tabelle), ist die Kinetik der Ringinversion des unsubstituierten Cycloheptans PR-spektroskopisch nicht messbar. Nach den Berechnungen von Hendrickson muss die Pseudorotation durch Methylgruppen jedoch sehr stark erschwert sein ($E_a > 9$ kcal/mole) (*4*). Bei methylsubstituierten Cycloheptanen sollten die konformativen Umwandlungsprozesse daher PR-spektroskopisch zu verfolgen sein. Wir haben deshalb eine Reihe von geeignet substituierten Cycloheptan-Derivaten untersucht (*5*). Hier sollen nur die Spektren von 5,5-Dimethyl-3,3,7,7-tetradeuterium-1,2-dithiacycloheptan diskutiert werden mit dem Ziel die Kinetik der konformativen Umwandlungen zu klären.

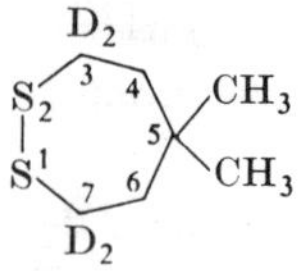

Das PR-Spektrum zeigt bei +25°C zwei einfache Signale bei $\tau = 8{,}18$ und 9,01 ppm (siehe Abb. 5), welche zueinander im angenähernden Intensitätsverhältnis 2:3 stehen. Das durch H-D-Kopplung geringfügig verbreiterte Signal bei $\tau = 8{,}18$ ist den 4,6-ständigen Methylenprotonen, das Signal bei

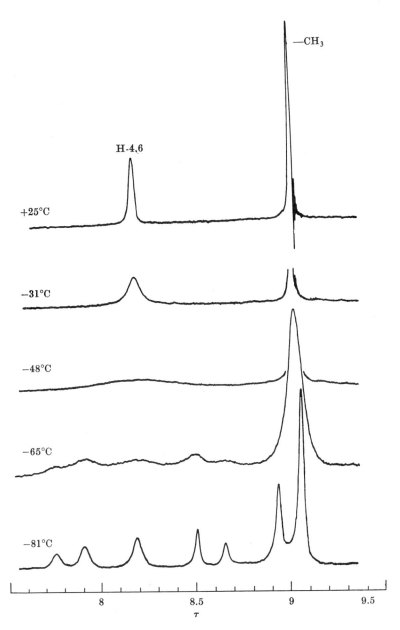

Abbildung 5 PR-Spektren des 5,5-Dimethyl-3,3,7,7-tetradeuterium-1,2-dithiacyclohep-
tans, aufgenommen bei einer Messfrequenz von 100 MHz im Temperatur-
bereich von +25 bis −81°C (in CS_2).

$\tau = 9{,}01$ ppm den Methylprotonen zuzuordnen. Beim Abkühlen der Probe werden beide Signale zunächst verbreitert und dann aufgespalten. Das Methylensignal geht bei $-57°C$ in ein AB-Quartett ($\tau_A = 7{,}81$; $\tau_B = 8{,}55$ ppm, $J = 15{,}0$ Hz) und ein Singulett bei $\tau = 8{,}19$ ppm über. Das Methylsignal spaltet bei $-73°C$ in ein unsymmetrisches Dublett ($\tau = 8{,}93$ und $\tau = 9{,}05$) auf. Dieses entsteht durch Überlagerung eines Dubletts und eines Singuletts. Verwendet man CS_2 als Lösungsmittel, stehen das AB-Quartett und das Singulett der Methylenprotonen im Intensitätsverhältnis $69:31$ (bei $-81°C$).

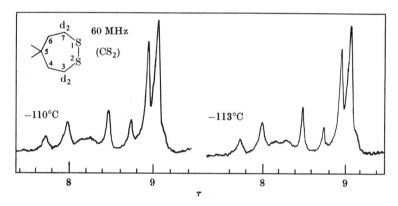

Abbildung 6 PR-Spektren des 5,5-Dimethyl-3,3,7,7-tetradeuterium-1,2-dithiacycloheptans, aufgenommen bei einer Messfrequenz von 60 MHz und Temperaturen von -110 und $-113°C$.

Bei weiterem Abkühlen der Probe wird auch das Singulett der Methylenprotonen bei $\tau = 8{,}19$ ppm verbreitert und bei $-110°C$ aufgespalten. Die Differenz der chemischen Verschiebungen wurde bei $-114°C$ zu etwa 18 Hz abgeschätzt.

ERGEBNISSE UND DISKUSSION

Bei Raumtemperatur erfolgen die konformativen Umwandlungen des Moleküls so schnell, dass im PR-Spektrum nur gemittelte Signale erscheinen. Beim Abkühlen der Probe werden nacheinander zwei verschiedene Umwandlungsprozesse eingefroren, die $\Delta G^{\neq}$-Werte wurden zu 10,8 und 8,0 kcal/mole bestimmt. Wie das Aufspaltungsbild des Methylensignals bei $\tau = 8{,}18$ ppm zeigt, wird bei $-57°C$ zunächst die gegenseitige Umwandlung zweier verschiedener Konformeren eingefroren. Von den im Spektrum bei $-81°C$ erscheinenden Linien ist das AB-Quartett der Methylen ($H_{4,6}$)-protonen und das Dublett der Methylprotonen dem Konformeren A, die beiden Singulette bei $\tau = 8{,}19$ und $9{,}05$ ppm dem zweiten Konformeren B zuzuordnen. Bei beiden Konformationen sind die Methylengruppen in 4- und 6-Stellung symmetrieäquivalent, die Molekel muss in diesen Konformationen ein entsprechend angeordnetes Symmetrieelement haben, d.h. es muss durch C-Atom 5 gehen. Im Konformeren A sind die Methylgruppen in 5-Stellung magnetisch nicht gleichwertig; bei diesem Konformeren kann es sich daher nur um C_{1v} und nicht um C_2-Symmetrie handeln. Beim Konformeren B kann man hingegen zwischen diesen beiden Symmetriearten nicht entscheiden, da das Methylsignal selbst bei den tiefsten Messtemperaturen (−114°C) nicht aufgespalten wird. Diese Befunde würde man nach den Erfahrungen an Cyclohexan- und Cycloheptenderivaten (6, 7) so interpretieren, dass das Konformere A die *starre* S_5- oder die W_5-Konformation besitzt. Diese Konformationen sind jedoch unwahrscheinlich, weil die Disulfidgruppe in der energiereichen *cis*-Anordnung vorliegen müsste. Das Konformere B könnte hingegen eine der symmetrischen Twistkonformationen TS_5 oder TW_5 einnehmen. Es lässt sich zeigen, dass das Auftreten getrennter PR-Signale für geminale Liganden bei gesättigten Siebenringverbindungen kein eindeutiges Kriterium dafür ist, dass die Molekel in der betreffenden Konformation tatsächlich starr ist. Aus dem Umwandlungsschema folgt (siehe Abb. 3), dass beispielsweise die gegenseitige Umwandlung der energiegleichen TS_{3+}- und TS_{7-}-Konformationen zu keiner formalen Vertauschung der axialen und äquatorialen Methylenprotonen und der Methylgruppen führt, wenn sie auf dem Wege

$$TS_{3+} \rightleftharpoons S_{6-} \rightleftharpoons TS_{2-} \rightleftharpoons S_{5+} \rightleftharpoons TS_{1+} \rightleftharpoons S_{4-} \rightleftharpoons TS_{7-}$$

verläuft. Auch bei sehr schneller gegenseitiger Umwandlung der beiden Twistkonformationen ist im PR-Spektrum für die Methylenprotonen ein AB-Quartett und für die Methylgruppen ein Dublett zu erwarten. Ausserdem würde diese Umwandlung eine durch Ringatom C-5 verlaufende Symmetrieebene vortäuschen. Anhand des PR-Spektrums kann also zwischen der *starren* S_{5+}- und den sich schnell ineinander umwandelnden TS_{3+}- und TS_{7-}-Konformationen nicht unterschieden werden.

Die diskutierten PR-Spektren unserer Verbindung erlauben verschiedene Interpretationen, von denen uns die folgende am plausibelsten erscheint: Das Konformere A gehört der Sesselfamilie an, oder das Konformere B gehört der Wannenfamilie an. Der langsamere der beiden experimentell beobachteten Umwandlungsprozesse ist mit dem energetisch günstigsten Versionsübergang zwischen beiden Konformationsfamilien identisch. Das bedeutet, dass der $\Delta G^{\neq}$-Wert eines Schrittes der Pseudorotation der Sesselfamilie mindestens so gross sein muss wie der des energetisch günstigsten Versionsüberganges zwischen beiden Konformationsfamilien. Dieser hohe Wert für eine Pseudorotationsbarriere stimmt mit der von Hendrickson (4) für Methylcycloheptan berechneten Pseudorotationsbarriere gut überein. Der schnellere der beiden beobachteten Umwandlungsprozesse ist sodann mit einem Teilschritt der Pseudorotation des Konformeren B identisch. Das heisst jedoch nach dem Vorhergesagten nicht, dass unter −110°C beide Konformere starr sein müssen, die innermolekulare Beweglichkeit durch eine partielle Pseudorotation kann durchaus eine *starre* Molekel vortäuschen.

LITERATURVERZEICHNIS

1. H.G. Schmid, A. Jaeschke, H. Friebolin, S. Kabuss und R. Mecke, *Org. Magn. Res.* **1**, 163 (1969).
2. J.B. Hendrickson, *J. Am. Chem. Soc.* **83**, 4537 (1961).
3. M. Bixon und S. Lifson, *Tetrahedron* **23**, 769 (1967).
4. J.B. Hendrickson, *J. Am. Chem. Soc.* **84**, 3355 (1962).
5. K. von Bredow, Dissertation, Freiburg i.Br. 1969.
6. H. Friebolin, H.G. Schmid, S. Kabuss und W. Faisst, *Org. Magn. Res.* **1**, 147 (1969).
7. S. Kabuss, H.G. Schmid, H. Friebolin und W. Faisst, *Org. Magn. Res.* **2**, 19 (1970).

Conformational Transmission in Steroids

Robert Bucourt

Centre de Recherches Roussel-Uclaf
Romainville, France

It is well known to the steroid chemist that an acid treatment converts the nonconjugated ketone, I, into its conjugated congener, II, quantitatively.

In sharp contrast, the 8-*iso* analog, III, shows a poor tendency towards conjugation. As we have recently stated (*1*), an acid treatment of compound III gives rise to an equilibrium mixture in which the starting product largely predominates. The product ratio is 2:1 of III to the conjugated material. In the case of the 8-*iso* series, the conjugated material is not unique; the two C-10 epimers are also present in equilibrium. Though this point is outside the present subject, our concern here being only the extent of conjugation at equilibrium, we feel we should mention this fact. There is a relative amount of about 9 parts of the compound having the α-orientated C-10 center, IV, for 1 part of the β-orientated compound, V.*

In an effort to understand these different trends to conjugation, we turned our attention to the bicyclic analog, VI. Earlier reports (*2, 3*) seem to indicate that the conjugation for compound VI also does not go to completion. The

---

* Whereas compound IV can adopt an all-chair and half-chair conformation, the ring B in compound V is forced to adopt a twist conformation. For this reason alone, V would not compete with IV. But the concomitant relief of much of the severe interactions occurring in IV, relatively lowers the energy of V to a point which allows its competition with IV.

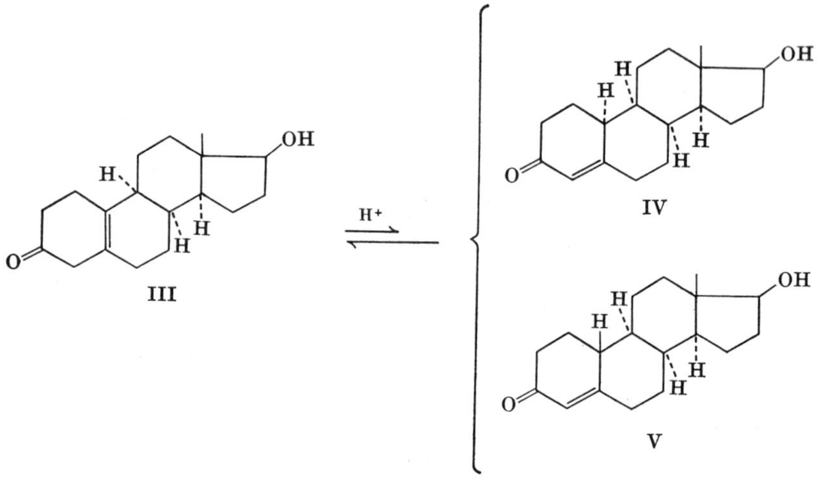

III   IV

V

study of the acid-induced equilibrium gave us a ratio of about 1:9 of the nonconjugated compound, VI, to the conjugated one, VII.

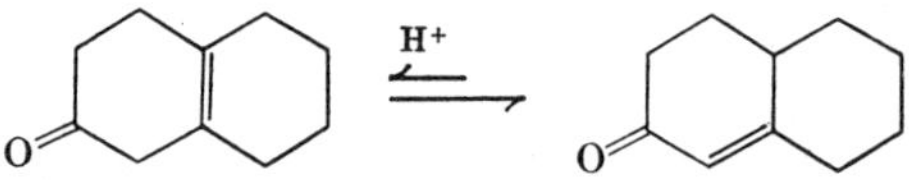

VI   VII

Thus, the stability relationships of the two steroid examples fall on each side of that of the decalone, VI. In the normal series, I, the presence of the additional C and D rings of the steroid framework enhances the conjugation, while in the 8-*iso* series, their presence hinders it. This result may be attributed to the intervention of a conformational transmission from the C ring to the B ring, acting upon the stability of the double bond in the 5,10 position.

It is the purpose of this communication to show that conformational analysis, by means of dihedral angles, can be helpful in the understanding of such a phenomenon. At first, it is necessary to recall the few basic principles underlying the method, which have already been published in our previous papers (*4, 5*).

As may be seen from Dreiding models, when two rings are *cis* fused, a variation of the dihedral angle of one nucleus at the junction gives rise to the same variation of the dihedral angle in the other nucleus at the junction. In accordance with the Newman projection, Fig. 1a, an opening gives rise to an opening, and a flattening to a flattening. On the contrary, the variation transmitted through a *trans* junction is inverted; an opening giving rise to a flattening (Fig. 1b), and a flattening to an opening.

The flattened chair form of cyclohexane having (internal) dihedral angles of 54° (Fig. 2a) is taken as a reference to compare and to characterize the

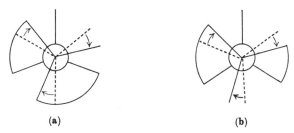

(a)   (b)

Figure 1   Newman projections.

geometries of the deformed rings. For example, a feature of the geometry of
the half-chair form of cyclohexene depicted in Fig. 2b is the deviation from
54° of its dihedral angles. The dihedral angles located 1,3 to the double bond,
having an absolute value of 44°, are said to be flattened (or closed) by 10° with
respect to cyclohexane (their absolute values are smaller than 54°). The dihedral
angle located 1,4 to the double bond, having a value of 61°, is said to be
opened by 7°.

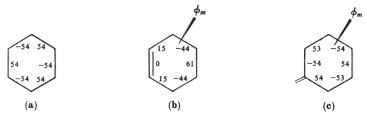

(a)   (b)   (c)

Figure 2

For cyclohexylidene (Fig. 2c), the geometry of minimum energy is not very
different from that of cyclohexane.

Returning to the phenomenon of conjugation, the two nuclei shown in
Figs. 2b and 2c can be considered as being a part of the steroid framework: the
B ring with the double bond in the 5,10 position for the former and the B
ring with the double bond in the 4,5 position for the latter. The value of the
dihedral angle of junction of the B ring with the C ring, marked ϕ_m on the
figures, being 44° on Fig. 2b and 54° on Fig. 2c, one may conclude that the
conjugation is associated with an increase of this angle. On the other hand,
the reverse transformation of the conjugated ketone into its nonconjugated
isomer is accompanied by a closing of ϕ_m.

In order to find out if these changes of ϕ_m agree or conflict with the state of
the C ring, we have now to examine the geometry of the steroid at this site.

The *trans* coupling of the C ring with the naturally flattened five-membered
D ring causes an opening of the dihedral angle of the C ring at this junction.
According to our calculations (5b), such an opening of a dihedral angle of a

cyclohexane chair form tends to flatten very slightly the dihedral angles located 1,3 to it. The C/B junction being located 1,3 to the C/D one, a value near 54°, and slightly lower, is expected for the dihedral angle of the C ring at the junction with the B ring. This reasoning has been recently supported by results of x-ray analysis (6).

In the case of the normal series (compounds I and II), the rings B and C being *trans* fused, an inverted variation is transmitted to the B ring. Thus, as a consequence of the above reasoning, the dihedral angle of the B ring at the junction with C must have a value near 54°, and slightly greater. This value fits well with the geometry of Fig. 2c, where this angle ϕ_m, has in fact the value of 54°. On the contrary, the value of 44° of Fig. 2b will conflict; while the effect of the C/B junction must be a slight opening of ϕ_m, the double bond in the 5,10 position on Fig. 2b should lead to a closing of ϕ_m. Therefore, we can conclude that the trend of steroid compound I to the conjugation must be greater than that of the bicyclic compound VI, which is in agreement with the experimental results.

The case of the 8-*iso* series (compounds III, IV and V) differs from the normal one, not only because the B/C junction is *cis*, but also because the 7-methylene axial to the C ring suffers a severe compression from the axial angular 18-methyl group. To relieve in part these interactions, the C ring has to flatten in the vicinity of the axial substituents (5c). Being rigidly held in the region of the 18-methyl group by the *trans* junction with D, the C ring will rather flatten at the side of the other axial substituent that is in the region of the flexible *cis* junction with B. Then the dihedral angle at this junction in ring C must suffer a notable closing. The same is true for the dihedral angle of junction in the B ring, on account of the *cis* fusion. As can be seen from Fig. 2b, the closing of 10° of ϕ_m corresponds well with the requirement of the C ring. So the position 5,10 for the double bond will be favored and the conjugation consequently hindered.

In summary, the distortion of ring C transmitted to ring B has a stabilizing effect upon the 5,10 double bond in the 8-*iso* series and a destabilizing one in the normal series.

REFERENCES

1. R. Bucourt, D. Hainaut, J.-C. Gasc and G. Nomine, *Tetrahedron Letters* **1968**, 5093; *Bull. Soc. Chim. Fr.* **1969**, 1920.
2. D.J. Baisted and J.S. Whitehurst, *J. Chem. Soc.* **1961**, 4089.
3. R.E. Juday, D.P. Page and G.A. Du Vall, *J. Med. Chem.* **7**, 519 (1964).
4. R. Bucourt, *Bull. Soc. Chim. Fr.* **1962**, 1983; **1963**, 1262; **1964**, 2080; **1967**, 1000.
5. R. Bucourt and D. Hainaut (a) *Compt. Rend. Acad. Sci.* **1964**, 3305; (b) *Bull. Soc. Chim. Fr.* **1965**, 1366; (c) **1966**, 501; (d) **1967**, 4562.
6. H.J. Geise, C. Altona and C. Romers, *Tetrahedron* **23**, 439 (1967).

Solvolytic Cyclizations Involving Double Bonds—Conformational Effects

Hugh Felkin
Institut de Chimie des Substances Naturelles, C.N.R.S.
Gif-sur-Yvette, France

"Organiser la participation là où elle ne l'est
pas encore, la développer là où elle existe,
voilà à quoi nous avons à nous appliquer."

CH. DE GAULLE (1968)

The solvolysis of suitably constituted olefinic arenesulfonates occurs with double-bond participation and leads to the (intramolecular) formation of a carbon–carbon bond. Thus, the solvolysis of the three unsaturated brosylates represented in Scheme I affords as major products the bicyclic compounds shown. Furthermore, the acetolysis rates are "accelerated" at 80°C with respect to the corresponding saturated brosylates by a factor of 63 in the case of cycloheptenylcarbinyl brosylate (*1*) and 4 in the case of cyclohexenylethyl brosylate (*2*), and the enol ether brosylate undergoes "solvolysis" in acetonitrile buffered with triethylamine 100 times faster than cyclohexenylethyl brosylate (*3*).

The purpose of this communication is to discuss the predicted and observed effect, on the degree of double-bond participation in these three systems, of putting a methyl group in place of the hydrogen shown (Scheme I) at the future bridgehead position.

A rough, computerless, prediction of this effect can be made by looking in more detail at the ground states of these brosylates and at the transition states leading to the bicyclic products.

Consider first the cycloheptenylcarbinyl system I. The ground state of this consists of an equilibrium mixture of a number of different conformers (Scheme II). We assume that there are four major ring conformers, two in which the CH_2OBs group is "equatorial" (Ia and Id), and two in which it is "axial" (Ib and Ic), and that there are in each case two rotational conformers (e.g., Ia and Ia') with respect to the C–COBs bond. Replacing R = H by

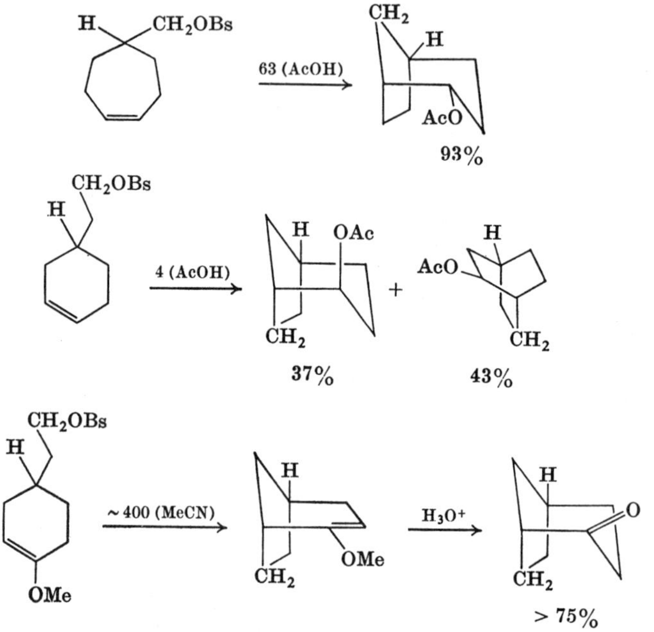

Scheme I

R = Me will introduce extra gauche C⟩C and C⟩O interactions (see Scheme III) into some of these eight major conformers (some of the CH$_2$ and OBs groups involved are indicated in Scheme II as ● and **O**Bs, respectively). Thus, in conformer Ia', when R = Me, there are two gauche butane (C⟩C) interactions

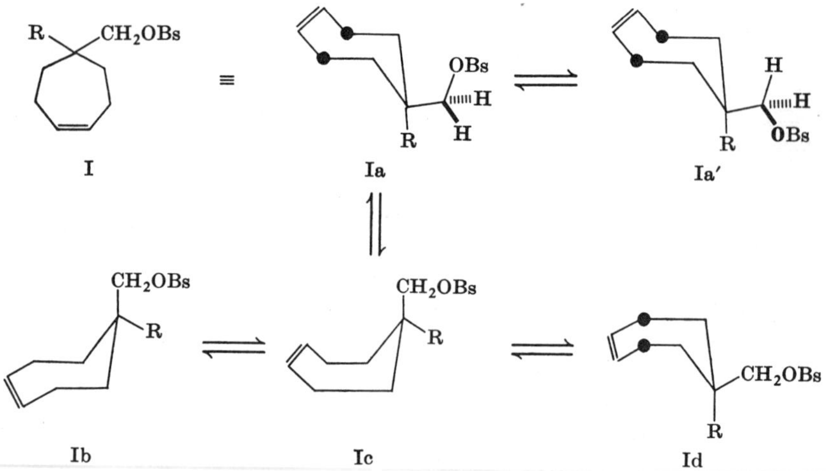

Scheme II

C $\overset{\chi}{}$ C = 0.85 kcal/mole   C $\overset{\chi}{}$ O = 0.4 kcal/mole

Scheme III

and one gauche C$\overset{\chi}{}$O interaction which do not exist when R = H. This extra strain in the ground state (ΔG_{calc}), due to the methyl group, can be roughly calculated (4), and amounts to about 1.2 kcal/mole on the basis of the values for gauche interactions shown in Scheme III.*

Now consider the transition states I$^+$ for the cyclization of these brosylates I (R = H and Me). These are represented in Scheme IV as nonclassical, leading

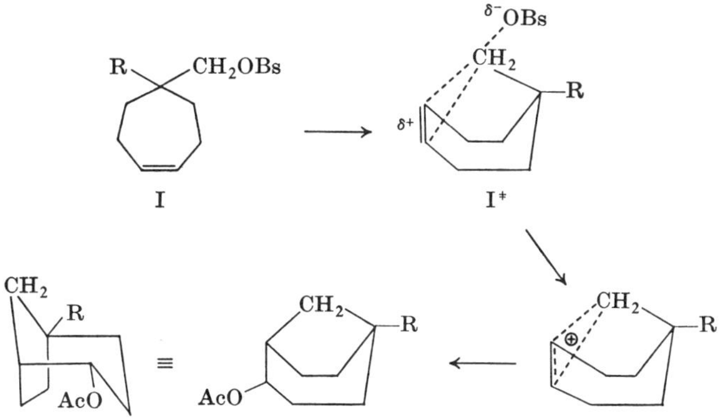

Scheme IV

to nonclassical bicyclic cations, but the argument would be the same with classical transition states and cations. If these transition states are assumed to be "product-like" (i.e., with a long C $\cdots$ OBs bond), the replacement of R = H by R = Me will introduce no extra strain, since there are no gauche butane interactions in I$^+$ involving R.

It follows that double-bond participation should be enhanced by about 1.2 kcal/mole in the methylated brosylate (I, R = Me) with respect to the unsubstituted brosylate (I, R = H). This is illustrated in Fig. 1, which shows

---

* It was pointed out in the discussion (by Professor Lüttringhaus) that the value of 0.4 kcal/mole for a gauche C$\overset{\chi}{}$O interaction may be high. The calculated magnitude (ΔG_{calc}) of the extra strain introduced by the methyl group is not, however, very sensitive to variations in this value. Thus, if 0.25 kcal/mole (=1/2 × 0.5 kcal/mole, the best A value for the OTs group: J.A. Hirsch, *Top. Stereochem.* **1**, 199 (1967)) is taken instead of 0.4 kcal/mole, ΔG_{calc} = 1.15 kcal/mole instead of 1.16 kcal/mole.

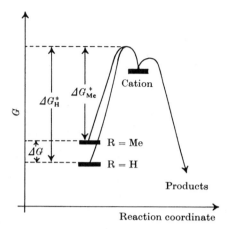

Figure 1 Energy profile for the solvolysis of the two brosylates I (R = H and Me).

the energy profile for the solvolysis of the two brosylates, drawn in such a way that the two transition states coincide. It is immediately apparent that $\Delta\Delta G^{*} (=\Delta G_{H}^{*} - \Delta G_{Me}^{*}) = \Delta G_{calc}$.

A similar reasoning can be applied to the two cyclohexenylethyl systems **II** and **III**. The four major ground-state conformers of these are shown in Scheme V, and, as in the case of the cycloheptenylcarbinyl system **I**, replacing R = H by R = Me will introduce extra gauche butane [Me⤹●] interactions into some of these conformers. This extra strain in the ground states (ΔG_{calc}),

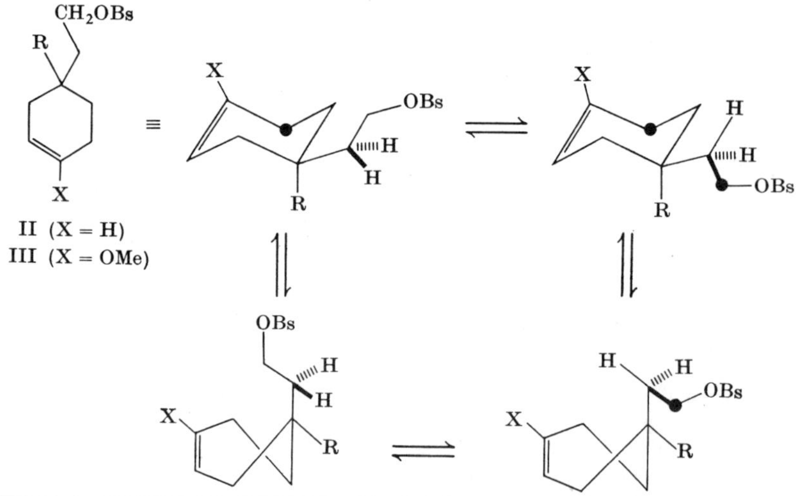

CH$_2$OBs
R

≡

X

OBs

$\overset{||||}{}$H
H

R

⇌

X

H

$\overset{||||}{}$H
OBs

R

II (X = H)
III (X = OMe)

⇅

⇅

OBs

$\overset{||||}{}$H
H

X

R

⇌

H   H
OBs

X

R

Scheme V

calculated in the same way (4), works out at about 1.9 kcal/mole. The transition states II^* for the cyclization of the olefinic brosylates II (R = H and Me) are shown in Scheme VI, and the transition states III^* for the cyclization

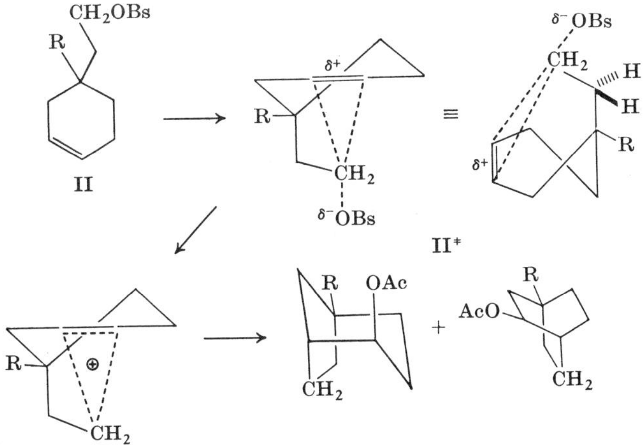

Scheme VI

of the brosylates with an enol ether group III (R = H and Me) in Scheme VII; II^* is represented as nonclassical and III^* as classical. It is apparent that, here again, the replacement of R = H by R = Me should introduce no extra strain in either of these transition states, so that double-bond participation is expected to be enhanced by 1.9 kcal/mole in both the methylated brosylates

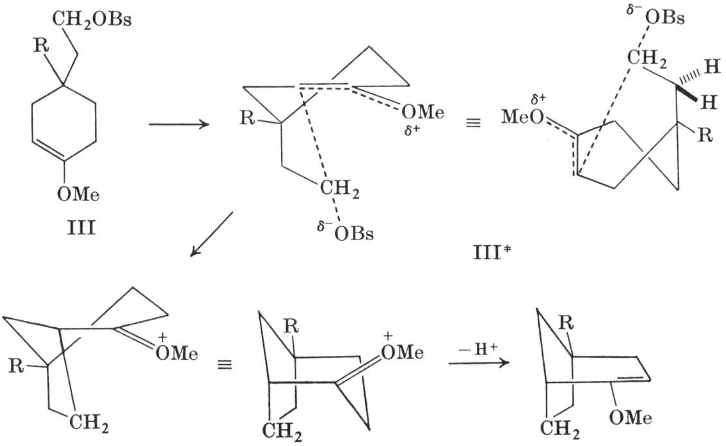

Scheme VII

II and III (R = Me) with respect to the unsubstituted brosylates II and III (R = H).

The three brosylates I, II and III (R = Me) have been prepared by unambiguous methods, and solvolyzed under the same conditions as the unsubstituted brosylates (R = H). In all three cases, the major products are the expected bicyclic compounds (R = Me) shown in Schemes IV, VI and VII. As regards rates, the two cyclohexenylethyl brosylates II and III are indeed more reactive with R = Me than with R = H, but the cycloheptenylcarbinyl brosylate I (R = Me) is much less reactive than the unsubstituted brosylate I (R = H). The figures are given in Table I.

Table I   Relative Rates of Solvolysis (k_{Me}/k_H) of the Brosylates I, II and III (R = Me versus R = H) and a Comparison between Calculated ($\Delta\Delta G^{*}_{calc}$) and Found ($\Delta\Delta G^{*}$) Values for the Differences in Free Energy of Activation ($\Delta G^{*}_H - \Delta G^{*}_{Me}$) Due to the Methyl Group

	I	II	III
k_{Me}/k_H			
AcOH, 80°C	0.02	25	
MeCN, 80°C		24	
MeCN, 40°C			42
$\Delta\Delta G^{*}$ (kcal)	−2.7	2.2	2.3
$\Delta\Delta G^{*}_{calc}$ (kcal)	1.2	1.9	1.9
Difference (kcal)	−3.9	0.3	0.4

It was stated in the discussion (by Professor Schleyer) that the formation of the bicyclic acetate (R = Me) shown in Scheme IV as the major product from the brosylate I (R = Me) was inconsistent with the fact that the acetolysis of this brosylate is 50 times slower than that of the unsubstituted brosylate I (R = H). It is difficult to see why this should be so. The brosylate I (R = Me) undergoes acetolysis at 80°C about 14 times faster than neopentyl brosylate, so that, if the Me and ring CH$_2$ groups in I (R = Me) behaved like the Me groups in neopentyl brosylate, I (R = Me) would afford about 7% (1/14) of products resulting from methyl migration and ring expansion, and 93% (13/14) of products resulting from double-bond participation. In fact, the propensity for migration of the Me and ring CH_2 groups in I (R = Me) appears (perhaps not unexpectedly) to be somewhat greater than that of the Me groups in neopentyl brosylate. The mixture obtained from I (R = Me) contains about 75% of products resulting from double-bond participation (67% of the bicyclic acetate (R = Me) shown in Scheme IV and 8% of the two isomeric bicyclic acetates (R = Me) shown in Scheme VI); the remainder of the mixture (about 25%, consisting of the two epimers of 4-acetoxy 1-methyl bicyclo[3.3.0]octane (3%), an unidentified acetate (2%) and unidentified hydrocarbons (20%)) results (we assume) from methyl migration and ring expansion.

The rate ratio (k_{Me}/k_H) at 80°C for the brosylates II is almost the same (25 and 24, respectively) in acetic acid buffered with sodium acetate and in acetonitrile buffered with triethylamine. This rate ratio corresponds to a difference in free energy of activation of 2.2 kcal/mole, which is in excellent

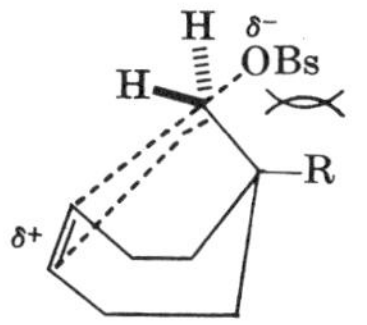

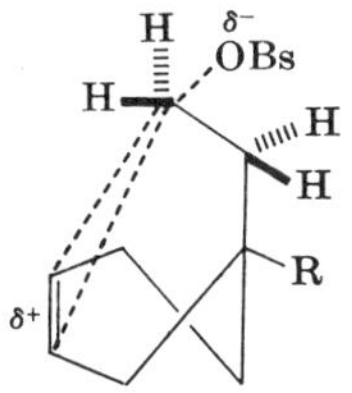

Scheme VIII

agreement with the calculated figure of 1.9 kcal/mole. Part of the difference (0.3 kcal/mole) could possibly be attributed to the inductive effect of the methyl group, since no account was taken of this in the rough calculations leading to the figure of 1.9 kcal/mole. Similarly, the rate ratio of 42 at 40°C for the brosylates III (R = Me and H) corresponds to a free-energy difference of 2.3 kcal/mole, which differs by only 0.4 kcal/mole from the calculated value of 1.9 kcal/mole.

The cycloheptenylcarbinyl brosylate I (R = Me), however, which was predicted to be about 5 times more reactive at 80°C (corresponding to a difference in free energy of activation of 1.2 kcal/mole) than the unsubstituted brosylate I (R = H), turns out in fact to be 50 times less reactive. The discrepancy in rates (a factor of 250) corresponds to a discrepancy in the calculated free energy of activation $(\Delta\Delta G^{\ddagger}_{calc})$ of 3.9 kcal/mole.

This is a very large figure, and we think it can only be due to steric hindrance to ionization in the substituted brosylate I (R = Me); this is illustrated in Scheme VIII. In the case of the cyclohexenylethyl compounds II and III, for which the agreement with calculation is very good, both the R and OBs groups are perfectly staggered in the transition state (e.g., II$^+$). This is not so, however, in the case of the cycloheptenylcarbinyl brosylate I. In the transition states I$^+$, the departing OBs group and the R group (H or Me) are eclipsed (which means there is extra strain in the transition state when R = Me). No account was taken in our calculations of any possible extra strain due to this, since it was assumed that, in a productlike transition state, the departing OBs group would be far enough away for any interaction between OBs and R to be negligible. It now seems that this was a very bad approximation.

The magnitude of the discrepancy in the calculated free energy of activation (3.9 kcal/mole) suggests, moreover, that in these transition states, the bond *angles* may have changed more, with respect to the ground states, than have the bond *lengths*. Two extreme possibilities for the transition states I$^+$ are shown in Scheme IX. In (a), which corresponds to our initial assumption, we

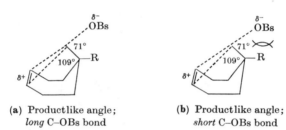

(a) Productlike angle; *long* C–OBs bond

(b) Productlike angle; *short* C–OBs bond

Scheme IX

have a productlike angle (180° − 109° = 71°) and a *long*, productlike, C···OBs bond, and this should not involve any significant extra strain on going from R = H to R = Me. In (b), we have a productlike angle (71°) and a *short*, ground statelike, C···OBs bond, so that the R and OBs groups will tend to be pushed together in the transition state, possibly even closer together than they would be in the eclipsed ground-state conformation. Such a transition state would therefore involve a large amount of extra strain on going from R = H to R = Me, and this would be consistent with the large value of 3.9 kcal/mole.

A somewhat similar idea has been put forward by Brown (5) to explain the relative slowness of the acetolysis of compounds such as *endo*-norbornyl tosylate; the departing OTs group on C-2 is assumed to be pushed towards C-6 in the transition state. If there is indeed such an effect in the *endo*-norbornyl system, it might be expected to be smaller than in our case, since the maximum angular change in an S_N1 reaction is about 19° (109° − 90°), whereas we are dealing with an "intramolecular S_N2" reaction in which the maximum angular change is much greater (109° − 71° = 38°).

ACKNOWLEDGMENTS

I thank my co-workers, C. Chuit, Mme. F. Colard and C. Lion, who did all the work, some of which has been published in preliminary form (*3, 6*). Acknowledgment is also made to the donors of the Petroleum Research Fund, administered by the American Chemical Society, for partial support of this research.

REFERENCES

1. G. Le Ny, *Compt. Rend.* **251**, 1526 (1960); H. Felkin, G. Le Ny, C. Lion, W.F.K. Macrosson, J. Martin and W. Parker, *Tetrahedron Letters* **1966**, 157.
2. S. Winstein and P. Carter, *J. Am. Chem. Soc.* **83**, 4485 (1961).
3. H. Felkin and C. Lion, *Chem. Commun.* **1968**, 60.
4. E.L. Eliel, N.L. Allinger, S.J. Angyal and G.A. Morrison, "Conformational Analysis," p. 23, Wiley (Interscience), New York, 1965.
5. H.C. Brown, I. Rothberg and D.L. Vander Jagt, *J. Am. Chem. Soc.* **89**, 6380 (1967).
6. C. Chuit, F. Colard and H. Felkin, *Chem. Commun.* **1966**, 118.

Conformational Aspects of N-Quaternization— a Combined Chemical and Nuclear Magnetic Resonance Study

Gabor Fodor
Department of Chemistry
West Virginia University
Morgantown, West Virginia

Nagabhushanam Mandava
United States Department of Agriculture
Beltsville, Maryland

Daniel Frehel and Mehru J. Cooper
Department of Chemistry
Laval University
Quebec, Canada

Fifteen years ago an unexpected selectivity in N-quaternization of tropanols was reported by Fodor and his associates (*1*). For instance, $3\alpha,6\beta$-tropandiol and ethyl iodoacetate ("direct" quaternization) led to the isolation of a single ester salt (I) which had two melting points (*2*); the second one proved to be that of the corresponding lactone (II). Reversing the sequence of quaternization, i.e., attaching the methyl group to the nitrogen in the last step, gave rise to the N-epimeric ester salt (III), which was unable to undergo cyclization (*2*) (Scheme I). These facts were in harmony with the concept of preferential equatorial quaternization. N-Methylation of N-ethyl nortropine afforded a tropanium salt as a main product, which was shown by x-ray crystallography to have an equatorial methyl (*3*). Furthermore, oscine (*2*) gave a lactone in the direct sequence of quaternization, while ecgoninol (*4*) did so in the reverse sequence. Since, however, the yields in lactonization were satisfactory only with the tropanols bearing an oxygen function in the pyrrolidine moiety, a certain risk as to the assignments was taken in those cases where the isolated lactone salt may have been the minor product of quaternization.

The equatorial course of quaternizations had been interpreted on conformational terms (*5*).

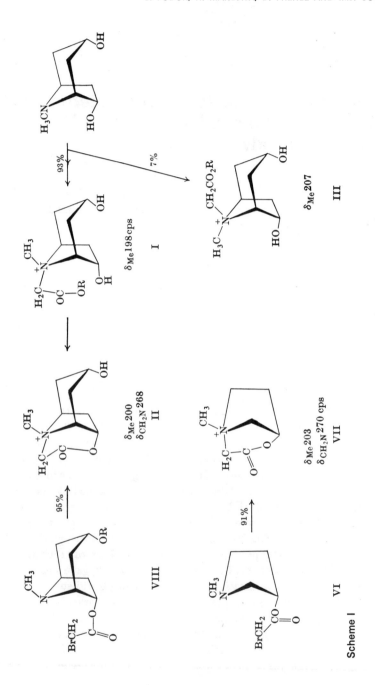

Scheme I

The use of nuclear magnetic resonance (nmr) in this field (first by Closs (*6*)) indicated that in the deuterohalides of the tropanes, the *N*-methyl group assumes two nonequivalent positions. Based on analogy with alkyl cyclohexanes, the lower-field methyl signal was assigned to the axial group.

Closs has inferred from these data that there is a kinetic rather than a thermodynamic control in quaternizations. J. McKenna and his pupils (*7*) have made a different approach to the same problem. In addition to empirical infrared (ir) criteria of less than general value, they compared the degree of stereoselectivity in the *direct* and *reverse* quaternizations with methyl and a higher alkyl halide. Based on isolated ratios of isolated products, these data were later completed

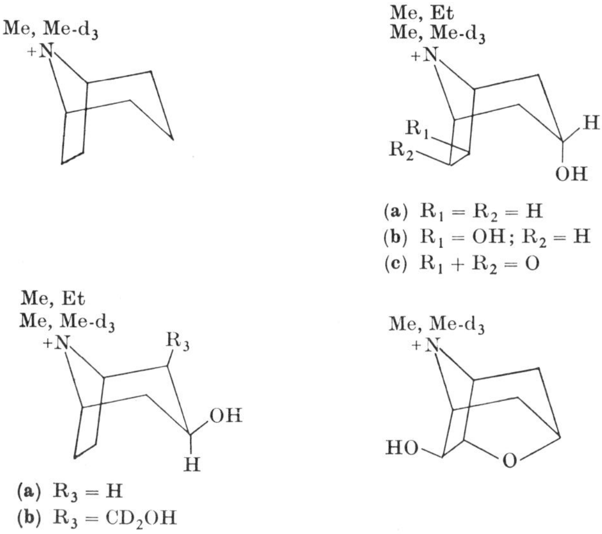

Scheme II

by nmr analysis. Comparative dequaternizations with thiophenoxide of *N*-epimeric heterocyclic ammonium salts have also been used (*8*). Very extensive work was focused on piperidines, camphidines and simple tropanes, with the conclusion that axial quaternization should be, and in many cases actually was, preferential. This series of papers (*7*, *8*) prompted us, two years ago, to have a second look at the field which we had investigated earlier with limitations imposed by classical methods, but this time, we were using more quantitative methods than before, in particular, nmr.

Product analysis of a series of tropanes (Scheme II), submitted to deuteromethylation in the nmr tube, had been carried out by taking the *N*-methyl signals in *N*,*N*-dimethyl nortropanium salts as reference, since these are separated by 2–10 cps.

Interpretation (9) of these results (as given in Table I) was based on the assumption of the equatorial methyl being more deshielded than the axial one. This has meant that while the parent compound and its 3α- and 3β-hydroxy derivatives undergo axial quaternization preferentially, the 6 and/or 7 oxygenated tropanols give equatorially quaternized salts. 2β-Hydroxy and 2β-hydroxymethyl compounds IV and V showed a less selective steric course.

Table I   Product Analysis of the Deuteromethylation of Tropanes by Nuclear Magnetic Resonance in DMSO-d_6

Tropane	N-Me (equatorial)	N-Me (axial)	Steric course[a] (% by integration)
3α-Hydroxymethiodide	188.5	185.2	
methiodide-d_3	189	185	85 axial[a]
Methyl-d_3-normethiodide	187	182.5	83 axial[a]
3β-Hydroxymethiodide	194	184	84 axial[a]
methiodide-d_3	194	184	84 axial[a]
2β-Hydroxymethyl-d_2-3β-hydroxy-methiodide	192	184	
methiodide-d_3	198	184	62 axial[a]
2β-Hydroxymethiodide	204	184	
methiodide-d_3	204	184	62 axial[a]
3α-6β-Dihydroxymethiodide	199	186	
methiodide-d_3	202	188	87 equatorial
3α-Hydroxy-6β,7β-epoxymethiodide	202 (200)[b]	186 (184)[b]	
methiodide-d_3	202 (203)[b]	186 (184)[b]	90 equatorial (93 equatorial)[b]
3α,6β-Oxido-7β-hydroxymethiodide	203 (198)[c]	200 (182)[c]	
methiodide-d_3	202.5 (198)[c]	199 (182)[c]	74 equatorial

[a] In view of recent correlation work assignments, the first four entries in this column now have to be reversed; the fifth and sixth entries remain to be checked.
[b] In D_2O.
[c] In CD_3CN.

These interesting assignments (9) have been shaken somewhat by a fact we recognized (10) shortly thereafter, namely that a neighboring hydroxyl or epoxide group has a strong deshielding effect (7.5–15 cps) upon the N-methyl. If this O-function is close to an equatorial methyl, its effect should coincide with overall equatorial shielding, in the case of an axial methyl group, however, it may counteract with overall equatorial deshielding, thus making any assignment of configuration rather uncertain. This casts doubt, in particular, on the N-configuration of compounds IV and V with one and two hydroxyls, respectively, close to the axial methyl group (Scheme III).

It seemed desirable to make a more profound correlation of *N*-methyl signals with *N*-configurations by combining nmr data with chemical facts. In order to do this, we needed safe reference compounds to which chemical shifts could be assigned unambiguously. The "old" method of lactone salt formation (*1, 2, 4*), starting from *N* towards *O*, has been substituted by an alternative *O* to *N* approach which we published first (*11*) in 1964 in overbridging the pyrrolizidine ring of oxyheliotridane (which proved useful also in other fields (*12*)), and recently to IX. Scheme I shows this idea applied to the conversion of *O*-bromoacetyl 3-pyrrolidinol (VI) into the lactone salt (VII), then to bromoacetyl 3α,6β-tropandiol (VIII), giving rise to a lactone (II) on *internal* quaternization.

In this context, *direct* quaternization of tropandiol with both methyl and ethyl bromoacetates was followed by nmr, thus confirming the formation (93%) of ester (I) and lactone (II) (which are interconvertible) besides the

Scheme III

N-epimeric ester salt (7%), described earlier as the main product (III) of *reverse* quaternization. Thus, an equatorial approach was confirmed once more for this case (Scheme I). Also, lactone formation by internal alkylation gave good yields.

This second method, *internal* quaternization, was applied subsequently to pseudotropine *O*-bromoacetate, independently of and simultaneously with us (*14*) by Bottini of the University of California, Davis (*13*). Liberating the ester base and heating in acetonitrile (or ether), a probable quaternary salt (XI in Scheme IV) precipitated in a high yield.

Ring closure of pseudotropine *N*-acetic acid previously (*1*) gave very poor yields of this lactone salt. Hydrolysis of the lactone then led to the same betaine and *N*-acetic acid as *direct* quaternization of pseudotropine (XIII).

However, checking this "lactone salt" by osmometry, we have found* extremely high apparent molecular weights, ranging from 513 to 520. On the other hand, the mass spectrum taken with Varian Model M-66 showed the molecular ion M^+ 181 and 183, besides a species which indicated incorporation

---

* Measurements were taken by M.J. Cooper at Laval University and later by H. Seguin at the National Research Council in Ottawa.

of the bromine (m/e 261 and 263) and also the MBr minus H, but no higher masses. This discrepancy might be explained either by assuming solvent association in the osmometric measurement or, alternatively, that the mass spectrum is limited to volatile fragments. Therefore, the "lactone" may be formed, indeed, by intramolecular cyclization (X → XI), justifying the con-

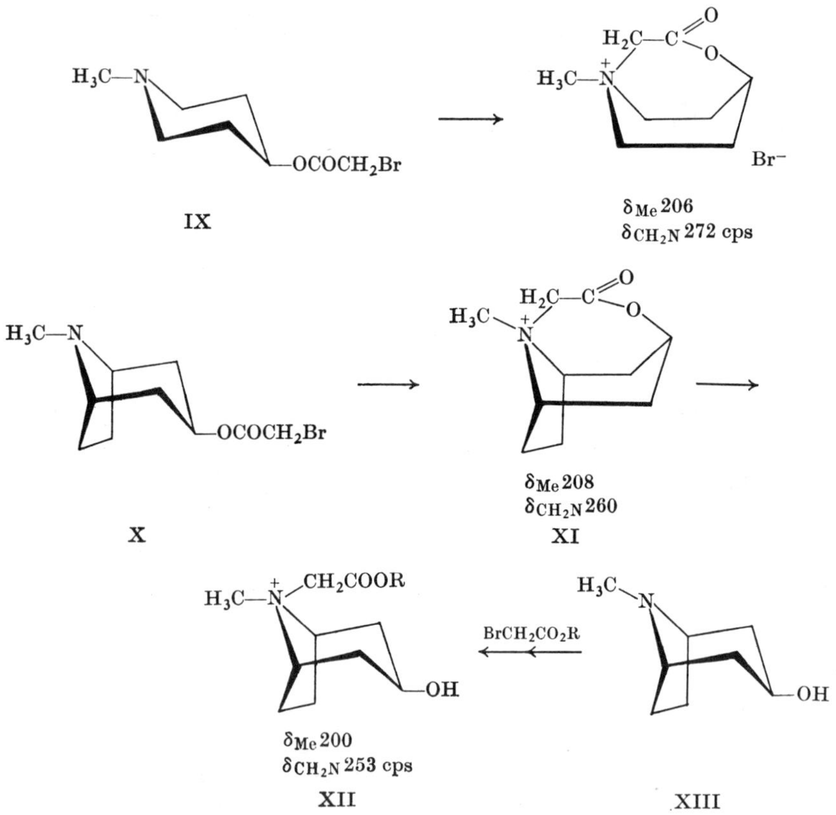

Scheme IV

clusions drawn as to axial quaternization of pseudotropine in view of the reaction XI → XII. *Inter*molecular, hence nonstereoselective, quaternization of bromoacetyl pseudotropine to give a macrocyclic lactone salt could not be completely excluded either. This second alternative may obviate any configurational conclusion of quaternization. At this stage, Chastain* was asked

* R.V. Chastain at West Virginia University.

to investigate this salt by x-ray diffractometry. Unfortunately, the inappropriate crystallographic properties of the salt have prevented, so far, any profound analysis. The *N*-carboxymethyl salt, however, is being investigated.*

It was thus deemed important to create a correlation between *N*-ethyl nortropine methobromide, which had an equatorial methyl group (XVIII), and the carboxymethylated tropine and pseudotropine.

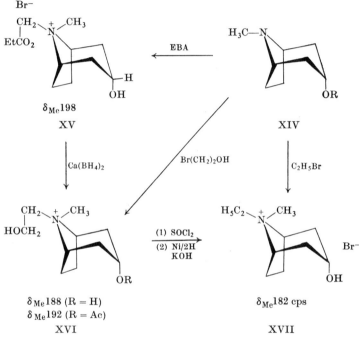

Scheme V

The formulas in Scheme V show that after a little detour—we had to protect the 3α-acetyl group before exchanging hydroxyl by chlorine of the 2-hydroxyethyl group—we finally obtained the *N*-ethyl tropanolium bromide (XVII). This *N*-ethyl derivative in the nmr had *N*-methyl signal at 182 cps (in DMSO) and the quartet of *N*-CH$_2$ centered at 205 cps (Fig. 1a), while the *N*-epimer, which was subjected previously to x-ray analysis, had *N*-Me at 180 cps and *N*-CH$_2$ centered at 210 cps (Fig. 2). Double irradiation of the methyl at −124.5

* Note added in proof: Chastain and Method have proven since that the sodium *N*-carboxymethyl 3β-hydroxytropanium bromide has an equatorial CH$_2$COONa group thus invalidating structure XII of the acid. By inference all configurational conclusions that were based on the "lactone" as being XI were incorrect, for it is a polymer unlike the product XI described earlier (*1*). x-Ray data will be submitted in full, soon for publication to *Acta Crystallographica*.

and −129 cps upfield, respectively, made the quartet collapse in both cases to a singlet at 205 and 211 cps (Figs. 1a and 2). This ethobromide has also been identified by an x-ray powder diagram with tropine ethobromide (XVII).*

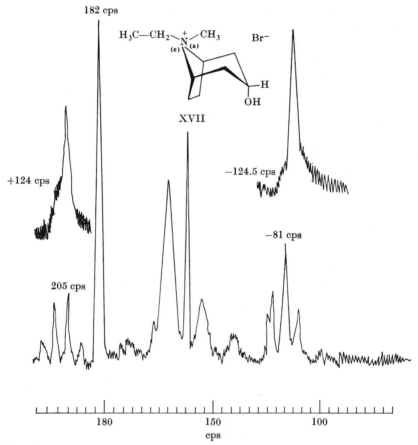

Figure 1a

The chemical interconversion showed that the major products of direct *N*-carbethoxymethylation (XV), *N*-2-hydroxyethylation (XVI) and *N*-ethylation (XVII) of 3α-tropanol (XIV) all had formed by the same steric course.

* Note added in proof: *N*-Ethylnortropine methobromide was earlier subjected to investigation by x-ray crystallography (*3*) while tropine ethobromide crystals had been affected by x rays. Now MacGillavry, Stam and Benci (to be published) have succeeded in solving the problem by collecting data at liquid air temperature. They kindly submitted the photo included here as Fig. 1b, proving that the ethyl is equatorial.

These interconversions can only be depicted by Scheme V, leading to the conclusion of a preferentially equatorial attack in quaternizations of tropanols. These facts are in apparent contradiction with the conclusion drawn from intramolecular alkylation of bromoacetyl pseudotropine to a lactone (Scheme IV), unless tropine (XIV) and pseudotropine (XIII) had reacted in an opposite way in quaternizations. In order to prove or disprove this possibility, an attempt was made to connect the two sets of alkylations via the ketone.

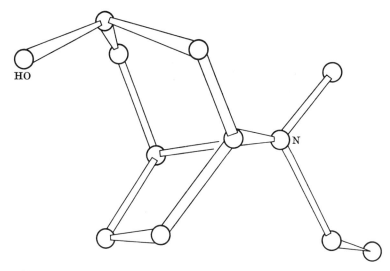

Figure 1b

Tropinone gave, on direct quaternization with ethyl bromoacetate, a quaternary salt as the main product (XIXa or XXa) with *N*-methyl signal at 218 cps (in D_2O). Catalytic hydrogenation of the latter afforded the bromide (XV), which, according to nmr data, was identical with the one from direct carbethoxymethylation of tropine. Accordingly, quaternization of the ketone and of the axial alcohol took the same steric course.

However, it was more difficult to find a stereoselective method in reducing the quaternized ketone (XXa) to the *equatorial* alcohol. After several unsuccessful attempts, *inter alia* with sodium borohydride, we found that long reaction time with Al-isopropoxide in isopropanol gave good results. This main product (with equatorial OH) had an *N*-methyl chemical shift of 187 cps, very different from that of the product of direct quaternization of XIII (δ *N*-Me 202 cps). It proved to be identical with the product of *reverse* quaternization of pseudotropine. Therefore, either tropine is quaternized in a different way than pseudotropine, or there was epimerization on the ring nitrogen (XIXa ⇌ XXa) prior to, during or after the Meerwein reduction. The latter assumptions have no analogy in the tropanes, though McKenna has found (7) equilibrations in some cases with other heterocycles.

In order to check this point, we tried first to epimerize pseudotropine N-acetic ester under Meerwein-Verley conditions. Within 48 hours a partial equilibration indeed occurred, for a second methyl signal at 187 cps appeared.

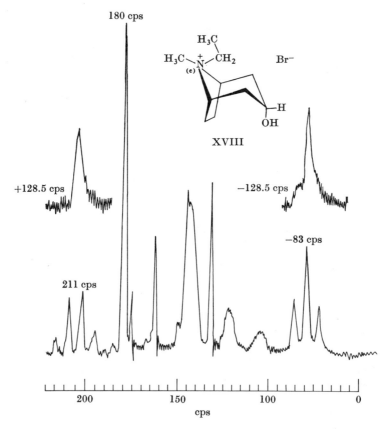

Figure 2

However, the rearrangement seemed to be too slow to account for the high yield of the product of the "anomalous" Meerwein reduction of the ketone. Therefore, the quaternized ketone (XXa) (δ N-Me 217 cps) was dissolved in t-BuOH and treated with Al-t-butoxide, giving no chance for reduction, while assuring the conditions for epimerization. Within a few hours at 35°C, a "new" N-Me signal actually appeared at 200 cps (Fig. 3). The same N-methyl signal was the predominant one when nortropinone was converted by *reverse* quaternization into the tropinonium salt (XIXa). Now as to interpretation of this interesting phenomenon, there were again three possibilities at hand: (1) dequaternization followed by requaternization—highly unlikely under Meerwein conditions; (2) Retro-Michael reaction, followed by rotation around the C-1–N bond in (XXI) and subsequent Michael addition to give the more stable *N*-epimer; (3) formation of a 3-azetidinone type of intermediate (XXII) followed by alcoholysis to either *N*-epimer XIXa or XXa.

The choice between the second and third alternatives (Scheme VI) could be found by using *N*-methyl-d$_3$ group and proving by nmr that the rearranged product contained the same as well as the CH$_2$CO$_2$R intact, or if an *N*-CH$_3$

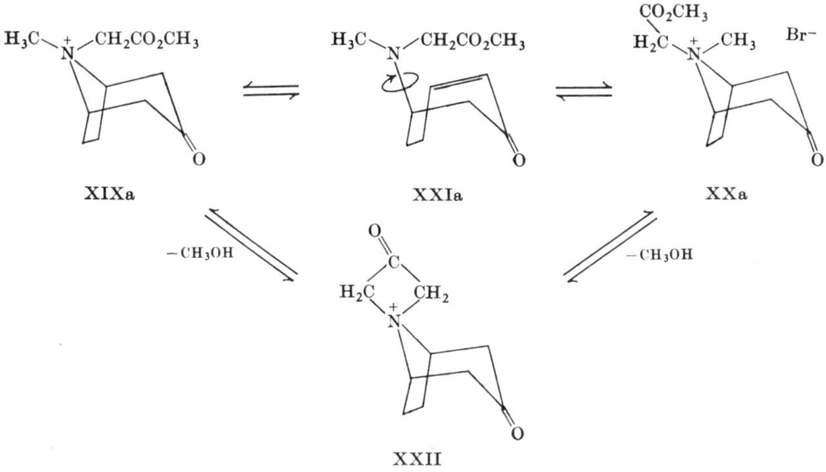

Scheme VI

group had formed. A preliminary experiment was done in *t*-BuOH with Al-*t*-butoxide with both CD$_3$-epimers XIXb and XXb. In order to eliminate overlap by *N*-CH$_2$ signal, the methyl esters were used instead of the ethyl esters. Indeed, a new *N*-CH$_3$ signal appeared in one case at 200 cps (Fig. 4); with the other epimer at 215 cps, exactly at the positions of the two non-deuterated *N*-methyl derivatives—in both cases at the expense of the methylene protons. This can best be reconciled with the concept of an azetidinone (XXII) type of transition state, except that the OCH$_3$ group was maintained. This difficulty could be overcome when assuming a tetrahedral azetidinediol hemiketal (XXV) as an intermediate formed from the less stable though more reactive ylide (XXIV) (Scheme VII). We have actually attempted to synthesize the spiro-3-azetidinone from nortropinone (XXVI) and 1,3-dibromoacetone, in two steps, and we will try to convert it into the dimethyl ketal. The behavior of the derivatives (there are but two 3-azetidin-ones reported (*15*, *16*) in the literature) towards alcoholysis has still to be studied.

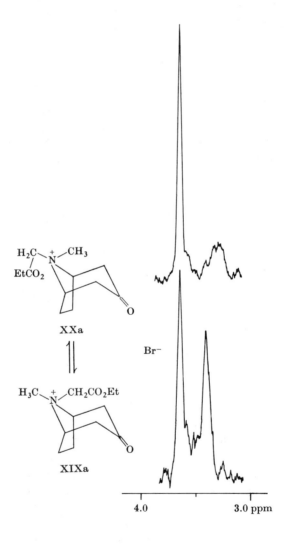

Figure 3

The objection which could be raised against this mechanism is the large difference in epimerization of the quaternized ketone versus the 3β-tropanolium derivative. This is more consistent with the fact that a quaternized amino ketone undergoes a Hofmann type of elimination with greater ease than an amino alcohol.* However, conformational effects also may enter into this picture, for the chair form of piperidine in tropinone is certainly much more flattened (XXVIII) than in pseudotropine (XXVII), although XIII and

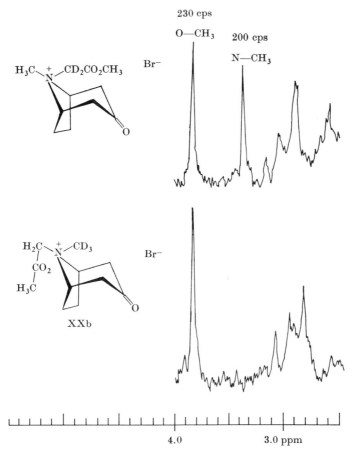

Figure 4

many other tropanols show coupling constants (*17*, *18*), pointing to dihedral angles H-2/H-3 larger than in the normal chair form of a piperidine (Scheme VIII).

* Note added in proof: Bottini kindly informed us that Nb-ethyl 3-tropanonium bromide underwent similar rapid interconversion in D_2O under the action of pyridine as a base into its *N*-stereoisomer. No transient cyclization is possible in that case, therefore, alternative 3 now seems to be corroborated.

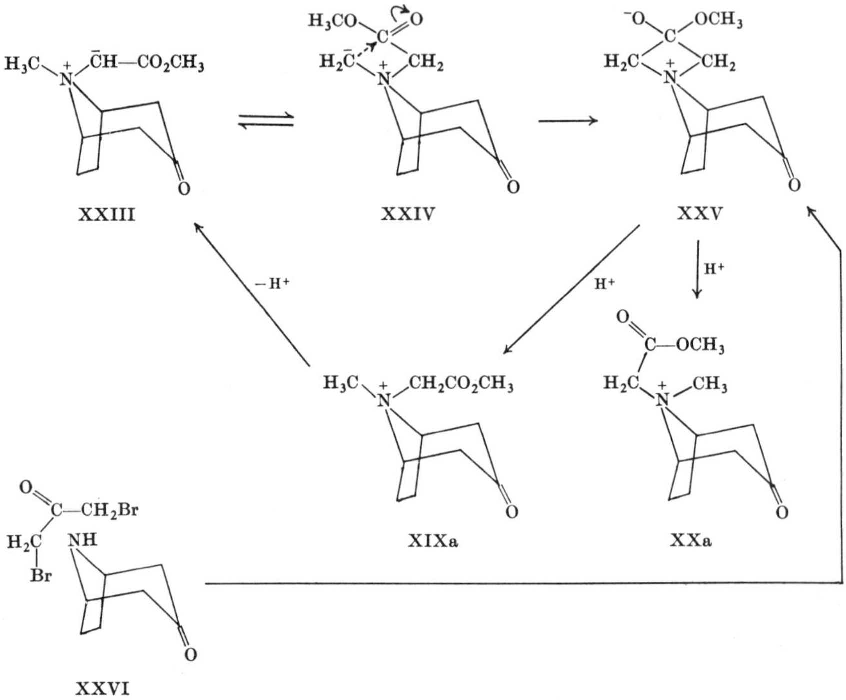

Scheme VII

Another factor may be the formation of a H-bridge between the partly enolized carbonyl and the axial substituent on nitrogen which may stabilize the more bulky group—but also may act as a repulsive factor. The ir data do indicate a sort of enolization, without providing, however, evidence for a hydrogen bonding.

This curious rearrangement is being investigated now by nmr, using the two nondeuterated and four partly deuterated N-methyl-N-methoxycarbonyl-methyl tropanonium bromides (Table II). For kinetic measurements, DMSO seems to be a better solvent for both the salts and Al-t-butoxide. However,

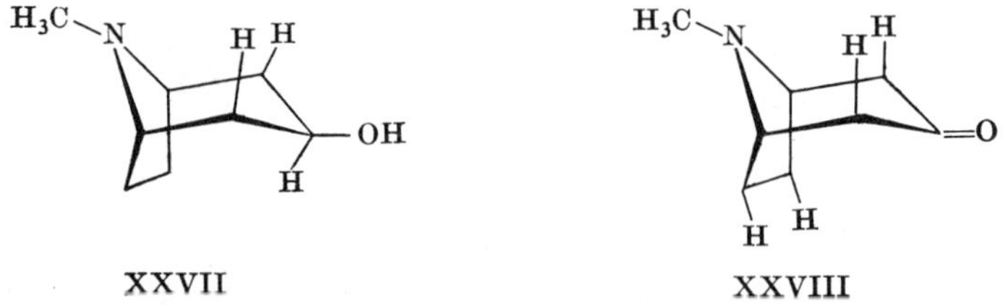

Scheme VIII

Table II Nuclear Magnetic Resonance Data (60 Mc) of the Carbomethoxy-methyl Tropinones (in cps)

N-Epimers	*N*-CH$_3$		*N*-CH$_2$		*O*-CH$_3$	
	D$_2$O	DMSO-d$_6$	D$_2$O	DMSO-d$_6$	D$_2$O	DMSO-d$_6$
H$_2$C–CO–OMe / CH$_3$ Br$^-$	215	214	264	277	230	229
H$_2$C–CO–OMe / CD$_3$ Br$^-$			261	280	230	230
D$_2$C–CO–OMe / CH$_3$ Br$^-$	214	213			231	229
H$_3$C / CH$_2$CO$_2$Me Br$^-$	200	200	285	300	230	230
D$_3$C / CH$_2$CO$_2$Me Br$^-$			284	303	230	230
H$_3$C / CD$_2$CO$_2$Me Br$^-$	201	201			231	230

there still remained a gap in our correlation scheme between carboxymethyl pseudotropine and the tropine derivative. To fill this (and also to overcome the complications arising from the epimerization of the keto ammonium salt we have discussed), N-carbomethoxymethyl tropinium bromide (XXVIII) and N-carbomethoxymethyl pseudotropinium bromide (XXIX) have both been "deoxygenated" via the corresponding 3-chlorotropanium bromides and catalytic hydrogenolysis of the latter (Scheme IX). There is unambiguous nmr

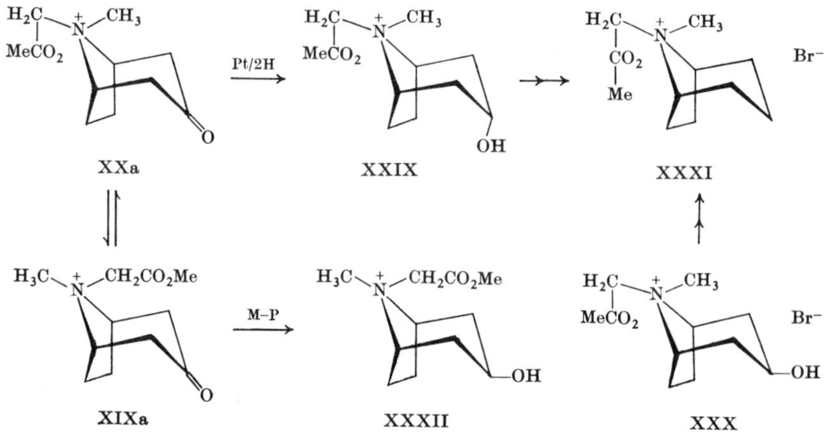

Scheme IX

evidence that the two N-carbomethoxymethyl tropanium bromides and also the major product in N-carbomethoxymethylating tropane are identical (XXXI) (N-Me resonates at 200 cps), and their whole nmr spectra are superimposable.*

Therefore, the final conclusions we can draw from these results are as follows: (1) Tropine, pseudotropine and tropinone are all quaternized with

Scheme X

* Bottini had kindly informed Fodor that K. Thut succeeded in oxidizing the major products of *direct* quaternization of tropine and of pseudotropine with ethylbromoacetate to the same ketone, itself arising from *direct* quaternization of tropinone.

bromoacetic esters following the same steric course. (2) Ethylation, carbalkoxy-methylation and 2-hydroxyethylation of tropine, i.e., alkylations with reagents of widely different bulkiness give products having the same configuration on nitrogen.

Accepting the previous x-ray data (*3*),* this means a preferentially equatorial course of quaternization of tropanes†—at variance with simple piperidines and camphidines (*7*).

Although this may seem surprising in view of previous criticism (*7*), we think the explanation Fodor gave previously (*19*) is satisfactory.

N-Alkyl groups point to a more favored equatorial position in the tertiary bases. In the transition state of quaternization, however, the axial alkyl conformation with an equatorial incoming cation seems more probable than the reverse, where the entering group is axially approached (Scheme X). On the basis of the Curtin-Hammett hypothesis, the free-energy differences between the two transition states should determine the ratio of the products. This concept must be true if the compression of the axial *N*-alkyl group by the C-2 and C-4 axial hydrogens is stronger than that between equatorial *N*-alkyl and the β-hydrogens in C-6 and C-7.

We wish to add one more piece of chemical evidence, i.e., conversion of the ester related to tropandiol lactone (II) of unambiguous structure into the same (or the *N*-epimeric) tropanium bromide (XXXI) via (XXXV), as the one from tropine and pseudotropine. Preliminary work on Scheme XI, which Fodor has presented in this chapter, led to a product with δ *N*-Me 206 cps, which ought to be further supported by analytical data.‡

However, the main question is now the apparent discrepancy between our product analysis in methylation-d₃ versus the consistent results in carbalkoxy-methylation and ethylation (Scheme V). Although it seems unlikely that methylation should go the opposite way, we still feel that chemical correlation as outlined on Scheme XII would still be desirable. Accordingly, a lactone (e.g., II) of proven structure should be converted into the *N*-CD$_2$COOH salt (XXXVI) followed by decarboxylation to give a deuteromethiodide of

---

* Refinement of the previous x-ray work on *N*-ethylnortropine methobromide and collection of new data with tropine ethobromide are in progress at the University of Amsterdam; see note added in proof on p. 80.

† Bottini let us know of the most recent x-ray work by Dr. U. deCamp on the product from pseudotropine and ethyl bromide, proving the equatorial orientation of the ethyl group therein. Although this compound had not been directly correlated with carbomethoxy-methyl pseudotropine, it is corroborating the view that tropanes undergo equatorial quaternization preponderantly.

‡ Note added in proof: This product was shown later to be the equatorial *N*-acetic acid and not its ester (XXXI). It was identified by using 100 Mc nmr with the products, prepared by deoxygenating of esters XXIX and XXX with subsequent hydrolysis to the *N*-acetic acid bromide, δ *N*-Me 3.34 and to the betaine, δ *N*-Me 3.24, as major products.

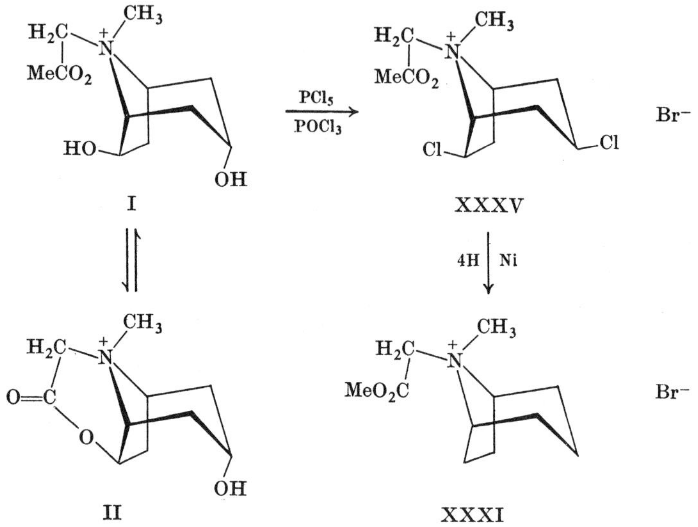

Scheme XI

unambiguous N-configuration (XXXVII). This work is already in progress.*

In any event, there are many inconsistencies in the assignment by analogy of low-field N-Me signals to equatorial configuration. For instance, products of direct N-carbethoxymethylation of tropine and tropinone, as we now know,

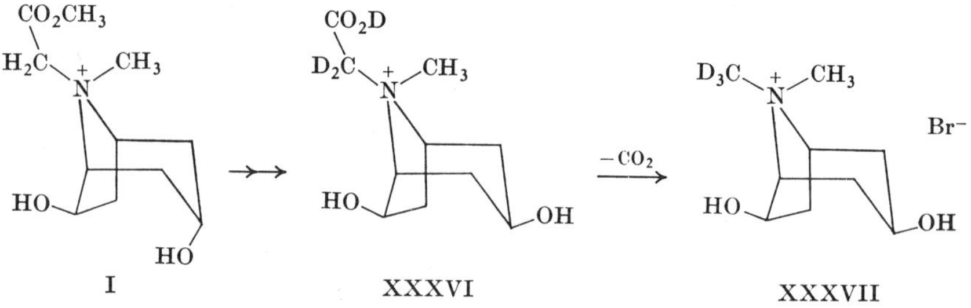

Scheme XII

* Note added in proof: These interconversions have now been achieved in two cases, i.e., with N-carbomethoxymethyl $3\alpha,6\beta$-dihydroxytropanium bromide (I) and with N-carbo-methoxymethyl 3α-hydroxytropanium bromide (XXIX). The N-deuteromethyl derivatives obtained in this way proved identical in both cases with the salts of type XXXVI we obtained on *direct* deuteromethylation from tropandiol and tropine, respectively. This means, that in tropandiol derivatives the equatorial methyl group resonates in the lower field, the opposite being true for tropane, tropine and pseudotropine derivatives. Therefore, configurational assignments for tropane, tropine and pseudo-tropine derivatives which we made in Table I ought to be reversed while 6- and/or 7-oxygenated tropanium salts were correctly interpreted.

all have axial methyl groups XVa and XXa with chemical shifts 6–15 cps, lower than the equatorial methyl in the "reverse" products, e.g., XIXa.

Similarly, the axial methyl in the methyl ethyl epimer resonates at 182 cps, while the equatorial one resonates at 180 cps. Among the tropanediol derivatives, however, the one with equatorial methyl has δ 207, while the axial is 198 cps.

In view of these facts, it should not be unexpected if the axial methyl in some deuteromethohalides would also resonate in a lower field. However, final conclusions cannot be reached before these last chemical correlations which we have planned are achieved (see footnote, p. 90).

Additional x-ray work (see footnotes, pp. 88–89) is also in progress.

Concerning the mechanism of *N*-epimerization, kinetic studies with deuterated compounds and also application of ^{13}C nmr spectroscopy are projected.

ACKNOWLEDGMENT

We wish to express our thanks to the Scientific Committee for the invitation to present this paper. Further, we would like to thank Drs. A.T. Bottini, C. MacGillavry, D. Arigoni, A.K. Bose, W. Houlihan, V.J. Traynelis, E. Grabowski and J. Sicher for helpful discussions, and the National Research Council of Canada for the grant which we worked under.

REFERENCES

1. G. Fodor, K. Koczka and J. Lestyan, *Magy. Kém. Foly.* **59**, 242 (1953); *J. Chem. Soc.* **1956**, 1411.
2. G. Fodor, J. Toth and I. Vincze, *Helv.* **37**, 907 (1959); *J. Chem. Soc.* **1955**, 3504.
3. C.H. MacGillavry and G. Fodor, *J. Chem. Soc.* **1964**, 597.
4. G. Fodor, O. Kovacs and M. Halmos, *J. Chem. Soc.* **1956**, 873.
5. G. Fodor, *Bull. Soc. Chim. Fr.* **1956**, 1032; *Alkaloids* **6**, 145 (1960).
6. G. Closs, *J. Am. Chem. Soc.* **81**, 5456 (1959).
7. D.R. Brown, R. Lygo, J. McKenna and J.M. McKenna, *J. Chem. Soc.* (*B*) **1967**, 1184 and references cited therein.
8. B.G. Hutley, J. McKenna and J.M. Stuart, *J. Chem. Soc.* (*B*) **1967**, 1199 and references cited therein.
9. G. Fodor, J.D. Medina and N. Mandava, *Chem. Comm.* **1968**, 581.
10. N. Mandava and G. Fodor, *Can. J. Chem.* **46**, 2671 (1968).
11. G. Fodor, F. Uresch, F. Dutka and T. Szell, *Coll. Czech. Chem. Comm.* **29**, 274 (1964).
12. H.O. House and associates, *J. Org. Chem.* **31**, 1002, 1073 (1966).
13. K. Thut and A.T. Bottini, *J. Am. Chem. Soc.* **90**, 4752 (1968).
14. N. Mandava, *Paper, 51st Annual Conference of the Canadian Institute of Chemistry, Vancouver, June 4, 1968*; reference *Congress Handbook*, p. 56.
15. C. Sandris and G. Ourisson, *Bull. Soc. Chim. Fr.* **1958**, 345.
16. Teng-Yueh Chen *et al.*, *Bull. Chem. Soc. Japan* **40**, 2398 (1967).
17. R.J. Bishop, G. Fodor, A.R. Katritzky *et al.*, *J. Chem. Soc.* (*C*) **1966**, 74.
18. H. Schenk, C.H. MacGillavry, S. Skolnik and J. Laan, *Acta Cryst.* **23**, 423 (1967).
19. G. Fodor, *in* R.H. Manske, *Alkaloids* **9**, 289 (1966).

Applicability of the Temperature-Dependences of Intensive Parameters for Quantitative Conformational Analysis— Conformational Analysis of Methylcyclohexane

Edgar W. Garbisch, Jr., Bruce L. Hawkins and Kenneth D. MacKay
Department of Chemistry
University of Minnesota
Minneapolis, Minnesota

Most molecules exist in one or more potential minimum conformations at ordinary temperatures, despite there being an infinite number of conformations accessible through internal rotations. An understanding of the origins of molecular conformational preferences requires at some stage the characterization of the populated conformers and the determination of their relative thermodynamic parameters. To accomplish this quantitatively for idealized systems containing but two conformers undergoing mobile conformational exchange is not always simple, and quantitative solutions of multiconformer (multidiastereomer) problems is markedly more difficult.

This paper discusses the scope and limitations of determining the relative thermodynamic parameters of conformers under mobile conformational equilibria using the temperature-dependences of averaged conformer intensive parameters. The conformational analysis of methylcyclohexane using the temperature-dependences of nuclear magnetic resonance (nmr) spectral parameters is given as an illustration of the approach.

I. GENERAL APPROACH

Consider the jth of l intensive parameters, P_{aj} and P_{bj}, of conformers A and B under mobile conformational equilibrium in solution. At the ith of k temperatures, the observed parameter, P_{ij}°, of the system

$$A \rightleftharpoons B$$

reflects an average of P_{aj} and P_{bj} weighted according to the relative populations of the two conformers, N_{aij} and N_{bij}. Provided no other conformers are
populated under the conditions of the parameter measurements,

$$P^\circ_{ij} = N_{aij} P_{aj} + N_{bij} P_{bj}$$

and

$$K_{ij} = \frac{N_{bij}}{N_{aij}} = \frac{P_{aj} - P^\circ_{ij}}{P^\circ_{ij} - P_{bj}}$$

or

$$\ln\left(\frac{P_{aj} - P^\circ_{ij}}{P^\circ_{ij} - P_{bj}}\right) = \frac{-\Delta H}{RT_{ij}} + \frac{\Delta S}{R} \tag{1}$$

for $j = 1, 2, \ldots, l$ and $i = 1, 2, \ldots, k$. With the requirements that (1) P_{aj}, P_{bj},
and ΔH be temperature-independent, so that the temperature-dependences
of P°_{ij} reflect the temperature-dependence of N_{aij} and N_{bij} at constant ΔH
and ΔS; (2) the conformers form an ideal solution; and (3) only two conformers
exist; equation (1) can be solved, in principle, for unknowns P_{aj}. P_{bj}, ΔH,
and ΔS, utilizing the temperature-dependences of P°_{ij}. Wood, Fickett and
Kirkwood (1) appear to be the first to recognize the solubility of equation (1)
from the temperature-dependences of P°_{ij} and the approach since has been
extensively pursued by others (2).

Various approaches have been employed in solving equation (1) for up to
$(2l + 2)$ unknowns (2). All, however, consider solution values of the unknowns
to be those that generate a minimum value of ϕ, as given by

$$\phi = \sum_{i=1}^{k} \sum_{j=1}^{l} (P^\circ_{ij} - P^{calc}_{ij})^2 \tag{2}$$

where P^{calc}_{ij} may be written as

$$P^{calc}_{ij} = \frac{P_{aj} - P_{bj}}{1 + \exp(-\Delta H/RT_{ij} + \Delta S/R)} + P_{bj} \tag{3}$$

after rearrangement of equation (1) and replacing P°_{ij} by P^{calc}_{ij}.

We adopted the iterative least-squares approach of Castellano and
Bothner-By for nmr spectral analysis (3) to determine values of the unknowns
that lead to a minimization of ϕ and hence to the solution of equation (1).

If the quantity $(P^\circ_{ij} - P^{calc}_{ij})$ is approximately linearly related to corrections
to the n unknowns, ΔU_s, according to

$$(P^\circ_{ij} - P^{calc}_{ij}) \approx \sum_{s=1}^{n} \frac{\partial P^{calc}_{ij}}{\partial U_s} \Delta U_s \tag{4}$$

for $i = 1, 2, \ldots, k$ and $j = 1, 2, \ldots, l$, a set of corrections, ΔU_s, may be obtained
by a method of least squares, which, after application to the initial values

of the unknowns, ^0U_s, provides new values for the unknowns, ^1U_s, and for P_{ij}^{calc} that will reduce ϕ (equation (2)). These new values of P_{ij}^{calc} are returned to equation (4) and the process repeated and continued until all values of ΔU_s are less than 0.001 units. Thus, a practical minimization of ϕ is realized.

CONAL 2 is a least-squares Fortran IV computer program that was written for calculating the best values of up to the $2l + 2$ unknowns (P_{aj}, P_{bj}, ΔH, and ΔS; $j = 1, 2, \ldots, l$). Additional details of CONAL 2 are given in the appendix to this chapter. A problem is initiated by computing the best least-squares values of P_{aj} and P_{bj} using equation (3), P_{ij}° (which replaces P_{ij}^{calc} in equation (3)), and initial estimates of ΔH and ΔS. These computed values of P_{aj} and P_{bj}, together with the initial values of ΔH and ΔS constitute the initial values of the unknowns to which corrections, ΔU_s, are applied in the first iteration.

As A or B may be the more stable conformer, there always are two solution values of ΔH and ΔS which differ only in sign. Also, the approach clearly does not characterize structurally the most stable conformer. Often, chemical intuition or the solution values of P_{aj} and P_{bj} will be sufficient for undisputed characterization.

The three above-mentioned requirements that led to the applicability of equation (1) have been discussed critically (2b, 4). It is difficult to generalize regarding the temperature-independence of conformer intensive parameters P_{aj} and P_{bj}. The best justification for assuming the temperature-independence of these parameters is to demonstrate the temperature-independence of similar parameters for molecules which are conformationally homogeneous or degenerate and structurally similar to the ones intended for conformational-analysis study. Since heat capacities of conformers normally will be similar (because of the general similarity of their vibrational spectra) and constant over the small temperature ranges adopted for ordinary variable temperature work, the difference between conformer heat capacities will approximate zero and the requirement of the temperature-independence of ΔH would appear to be met. However, for conformers having large differences in dipole moments, particularly in media of high dielectric constant, Abraham et al. (5) have demonstrated the likelihood of substantial temperature-dependences of ΔH. Should conformers strongly interact intermolecularly with themselves or the solvent, deviations from solution ideality will result and invalidate equation (1). Consequently, it is desirable to use inert solvents of minimum dielectric constant and to determine parameters, P_{ij}°, under conditions of high dilution when solute–solute interactions are probable. The requirement that only two diastereomeric conformers be significantly populated at the temperatures of the experiments limits practical application of the approach to relatively simple molecules.

II. SCOPE AND LIMITATIONS OF THE APPROACH

Routine testing of CONAL 2 using accurate synthetic data showed that convergencies to the correct solutions were independent of the initial guesses of ΔH and ΔS, provided that the guesses were within approximately 500 cal/mole and 3 eu of the correct values. Otherwise, convergencies were not realized.

To properly test the scope and limitations of the approach, synthetic problems were run using data that contained gaussian-distributed random errors. Thus, real data were considered to be approximately simulated. For these problems,

$$P_{ij}^{\text{syn}} = \sigma\mu + P_{ij}^{\text{calc}}$$

where P_{ij}^{calc} is given by equation (3), σ is the standard deviation in P_{ij}^{syn}, and μ is a random number from a gaussian distribution having a mean of zero and a variance of unity.

Using synthetic data containing random errors, we found that whenever CONAL 2 converged, it did so to give the same solution regardless of the initial guesses of ΔH and ΔS. However, with standard deviations in P_{ij}^{syn} of $>0.4y$, where

$$y = |P_{aj} - P_{bj}|(10^{-3})$$

equation (1) could not be solved for all the unknowns. That is, CONAL 2 rarely converged if all unknowns were varied, and when convergence was observed, the probable errors in ΔH and ΔS were intolerable. Because of the scatter in P_{ij}^{syn}, a shallow RMS surface is obtained, and essentially the same minimum value of ϕ is generated from many values of the unknowns. These results show that the approach cannot be applied reliably to solve problems using proton magnetic resonance (pmr) data where y generally is less than 0.02 cps, unless either ΔH or ΔS is held constant.

To solve problems using data having standard deviations, σ, ranging from $3y$ to $0.5y$, it is preferable generally to hold ΔS constant at a value consonant with the symmetries of the two conformers under equilibrium or at a value otherwise theoretically obtained. If parameters P_{aj} and P_{bj} of a conformer are known exactly or can be measured, then these parameters should be held constant at their correct values and the problem solved for the remaining unknowns.

We explored the scope and limitations in solving equation (1) with the entropy change fixed. Because of various experimental limitations, which include solvent and solute physical properties, the temperature range over which P_{ij}° is determined for a given sample is limited to roughly 200°. We pursued testing of the approach using two temperature ranges, namely −153 to 47°C (120–320°K) and −23 to 177°C (250–450°K), which subsequently we will refer to as the low- and high-temperature ranges, respectively.

It was found that the parameter probable errors in the solutions of equation (1) generally are smallest when P_{ij}° is measured over temperature ranges in which the change in P_{ij}° is greatest. So as to determine optimum experimental temperature ranges, curves were plotted showing the temperature-dependences of P_{ij}° for different values of ΔH, while fixing ΔS at zero. The results are shown in Fig. 1. The curves in Fig. 1 can be used reliably for systems having nonzero values of ΔS, as the temperature range which gives the largest change in P_{ij}° is insensitive to ΔS.

When ΔH is between 200 and 2000 cal/mole, it is desirable to determine P_{ij}° over the low-temperature range, and for values of ΔH above 2000 cal/mole the high-temperature range is preferable. Solutions of equation (1) for systems having ΔH below 200 and above 3000 cal/mole generally will have intolerably large probable errors associated with the solution parameters, as the standard deviations in P_{ij}° may approach the small changes in P_{ij}° versus temperature.

Consider an equilibrium having ΔH somewhere between 400 and 1000 cal/mole. The temperature-dependences of P_{ij}° in the high-temperature range are surprisingly similar over this range of ΔH (see Fig. 1). After superimposing a scatter in P_{ij}° due to experimental errors, it is easily understood why it will be difficult to obtain a reliable experimental estimate of ΔH. However, the temperature-dependences of P_{ij}° in the low-temperature range are quite different for these values of ΔH, and more reliable estimates of ΔH may be expected. For systems having ΔH above 2000 cal/mole, the temperature-dependence of P_{ij}° may be experimentally indeterminable in the low-temperature range due to measurement uncertainties, and the high-temperature range clearly is preferred. However, the temperature-dependences of P_{ij}° are modest and similar for equilibria having low ($\sim$400 cal/mole) and high ($\sim$2000 cal/mole) values of ΔH in the high-temperature range. This means that large probable errors will be associated with the solution parameters of equation (1), unless highly accurate data are obtained. In such instances, measuring P_{ij}° at several temperatures in the neighborhood of $-100°C$, if possible, would render a more reliable solution.

For practical purposes, it would be helpful to have numerical estimates of the expected probable errors in the calculated enthalpy change, ΔH_{calc} as a function of the accuracy of the observed parameters, P_{ij}°, and the correct ΔH over the high- and low-temperature ranges. Also, it is of interest to know whether ΔH is obtained significantly more reliably from one or more observed parameters (P_{ij}° where $j > 1$). Results related to these matters are presented graphically in Fig. 2.

The following salient conclusions may be drawn from the results shown in Fig. 2: (1) From data collected in the high-temperature range and having $\sigma = y$ to $3y$, the most reliable results are obtained for systems having ΔH

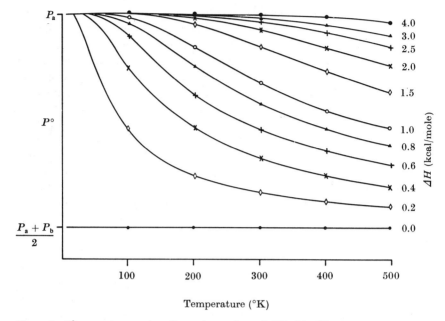

Figure 1 P^0 versus temperature for various values of ΔH with $\Delta S = 0$.

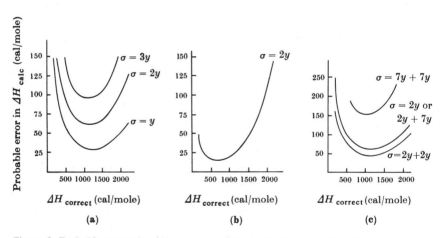

Figure 2 Probable errors in ΔH_{calc} versus $\Delta H_{correct}$ (cal/mole) using (a) one observed parameter obtained between -23 and $177°C$ and having the indicated standard deviations, σ; (b) one observed parameter obtained between -153 and $47°C$ and having $\sigma = 2y$; (c) one and two observed parameters measured between -23 and $177°C$ and having the indicated standard deviations (ΔS was fixed at zero).

between 600 and 1700 cal/mole. Using data having $\sigma = 2y$ and collected in the low-temperature range, the probable errors in ΔH_{calc} are acceptable for systems having ΔH between 200 and 2000 cal/mole. (2) The most accurate estimates of ΔH having correct values in the region of 300–2000 cal/mole are expected from data obtained in the low-temperature range. (3) The value of ΔH is slightly more accurately determined from two rather than from one ob-

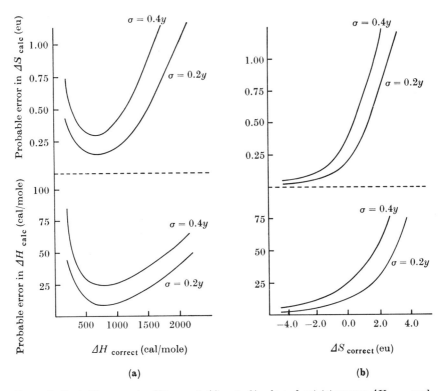

Figure 3   Probable errors in ΔH_{calc} and ΔS_{calc} (cal/mole and eu) (a) versus $\Delta H_{correct}$ and (b) versus $\Delta S_{correct}$, using one observed parameter obtained between -153 and $47°C$ and having $\sigma = 0.4y$ and $0.2y$. For (a) $\Delta S_{correct} = 0.0$ eu for all calculations. For (b) $\Delta H_{correct} = -1000$ cal/mole for all calculations.

served parameter having the same values of h in $\sigma = hy$. When two parameters are used, one having a large value and the other a small value for h, the probable error in ΔH_{calc} is governed by the parameter having the small value for h.

It was found that solutions of equation (1) for all unknowns could be achieved using parameters having $\sigma \leqslant 0.4y$ and obtained in the low-temperature range. Figs. 3a and b show some results of solutions of synthetic problems in the low-temperature range where ΔH and ΔS were varied. When $\Delta S_{correct} \approx$ 0 eu, reasonably accurate solutions were achieved for $\Delta H_{correct}$ of 300–1500 cal/mole when the standard deviations in σ were 0.2y and 0.4y. Also, it appears (Fig. 3b) that the most accurate solutions are obtained when ΔH and ΔS have the same sign.

We wished to determine whether or not the presence of more than two conformers might be detected by comparing the standard deviations in fitting the observed parameters, σ_{fit},* with those in P_{ij}°, σ_{exp}. Results of synthetic and real problems showed generally that σ_{fit} and σ_{exp} were of comparable magnitudes. The thought arose that treating multicomponent problems as two-component ones might lead to values of σ_{fit} that are significantly higher than those of σ_{exp}. This idea was tested. Synthetic values of P_{ij}° for three-component equilibrium problems were calculated using various relative enthalpies of the three components and various values for the component intensive parameters. These values of P_{ij}° then were assumed to arise from two components, and equation (1) was solved using CONAL 2. In all cases, σ_{fit} was greater than σ_{exp}, but not always significantly greater so as to permit general conclusions to be made. These findings, although preliminary, do not lead us to be optimistic toward either detecting the presence of more than two conformers or successfully solving multiconformer problems using the temperature-dependences of conformer intensive parameters.

III. CONFORMATIONAL ANALYSIS OF METHYLCYCLOHEXANE

We felt that it was necessary to test the described approach to conformational analysis on a two-conformer molecule for which ΔH and ΔS generally are considered known. Methylcyclohexane was our choice. There have been at least one theoretical (6) and twenty-seven experimental (7) determinations of ΔH, ΔS or ΔG for the conformational equilibrium (5),

and it appears that the generally considered best values of ΔH and ΔS for the liquid phase are −1700 cal/mole and 0.0 eu (6, 7). However, no experimental determinations of ΔH and ΔS for equation (5) have been made using methyl-cyclohexane.

The temperature-dependences of the two vicinal proton couplings and three

* $\sigma_{fit} = (\phi/m)^{1/2}$, where m is equal to the total number of observed parameters.

proton chemical shifts in I were used to solve equation (1) for ΔH and the ten unknown intensive parameters of the two conformers of I. The deuterium isotope effect on ΔH is expected to be negligible (8), so that the value of ΔH obtained may be considered that for undeuterated methylcyclohexane (equation (5)).

$$CD_3 \quad H_2$$

$$D_2 \qquad \qquad H_3$$

$$\qquad \qquad H_1$$

$$D_2 \qquad \qquad D_2$$

$$D_2$$

I

A typical deuterium decoupled pmr spectrum of I is shown in Fig. 4.* Spectra such as the one shown in Fig. 4, which are of the ABC type (9), were analyzed using LAOCOON3 (3). Of the two vicinal proton couplings, one is large (~11 cps) and the other small (~4 cps). The large coupling decreased markedly with increasing temperature and the small coupling was nearly temperature-independent. Consequently, the large coupling arises between the *trans* hydrogens in I, and because of its magnitude and its decrease with increasing temperature, the conformer bearing the equatorial methyl group clearly is the more stable of the two (see equation (5)).

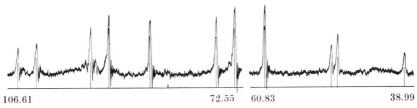

106.61   72.55   60.83   38.99

Figure 4   Deuterium decoupled pmr spectrum of I (top) and calculated stick spectrum (bottom). Numbers are transition frequencies downfield from TMS.

As discussed in Section I, equation (1) is applicable only if parameters P_{aj} and P_{bj} are temperature-independent. The observed (10) temperature-independence of the vicinal proton couplings in cyclohexane over 141°C, makes it reasonable to presume that these couplings in conformers A and B of equation (5) also are temperature-independent. However, the cyclohexane

* A 35% mole/mole sample of I in carbon tetrachloride containing 3% vol/vol of tetra-methylsilane was used. The sample tube was degassed and sealed at −68°C under 0.1 mm pressure. Eight to twelve sideband calibrated spectra were determined at each temperature (from −17 to 158°C). The frequencies of each observed transition in these spectra were averaged and the averaged values used for spectral analyses. Standard deviations in the experimental frequencies generally were less than 0.04 cps and probable errors in the computed proton couplings and chemical shifts were between 0.01 and 0.03 cps, and 0.02 and 0.04 cps, respectively.

proton chemical shift relative to TMS was found to change 0.5 cps (downfield) over a 90°C increase in temperature. As it is unknown whether the chemical shift of TMS is temperature-dependent or not, it will be necessary to demonstrate the compatibility of separate solutions of equation (1), using proton couplings and both intramolecular and intermolecular (TMS references) chemical shifts before a solution using the combined parameters can be considered reliable.

The temperature-dependences of the nmr parameters of I are shown in Figs. 5a and b. Solutions of equation (1) were not obtained by varying both

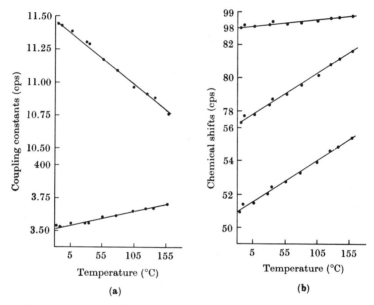

Figure 5   Temperature-dependences of (a) the vicinal proton coupling constants for I, (b) the proton chemical shifts for I.

ΔH and ΔS. This was expected, since y goes from about 0.01 to 0.06 as σ in the observed parameters, P_{ij}°, goes from about 0.02 to 0.04, respectively. It was necessary, therefore, to fix ΔS during solutions of equation (1). Holding ΔS constant at values between ±3 eu, solutions of equation (1) were obtained using the temperature-dependences of (a) the two vicinal coupling constants, (b) the chemical shifts of the three protons relative to TMS, (c) the two vicinal coupling constants and the chemical shifts of the three protons relative to TMS, (d) the intramolecular chemical shifts, $\delta_{1,3}$ and $\delta_{1,2}$, and (e) the intramolecular chemical shifts and the two vicinal coupling constants. The results of these calculations are collected in Table I. It is particularly noteworthy that the values of ΔH for solutions (a), (b), and (d) are essentially the same within the probable error limits for any fixed value of ΔS. And it is reasonable to conclude that the temperature-dependences of both the intermolecular and intramolecular chemical shifts are dominated by the temperature-dependence of the relative conformer populations. Solutions of the problem using both chemical-shift and coupling-constant data (solutions (c) and (e)) appear justified and the results therefrom should be regarded as being the most reliable. These results, however, are dominated by the temperature-dependences of the chemical-shift data where y is about six times that for the coupling-constant data and σ for both data are comparable (see Section II).

If $\Delta S = 0$, as generally accepted (6, 7), then a value of ΔH of -1655 ± 60 cal/mole, as given by the mean of solutions (c) and (e) of Table I, may be considered the best experimental value deriving from the temperature-dependences of the nmr parameters of I. The agreement between this value of ΔH and that (-1700 cal/mole) considered to be correct (6, 7), clearly is excellent.

Although a value of ΔS could not be obtained directly from the temperature-dependences of the nmr parameters of I, a rough estimate of the entropy change may be considered using equation (6). This equation gives the theoretical relationship (11) between the values of vicinal proton coupling constants and the dihedral angles, θ, between the two vicinal C–H bonds. Also, it is necessary to make the reasonable

$$J = A(\cos^2 \theta + X \cos \theta) \tag{6}$$

assumption that the constants, A and X, in equation (6) are the same for both conformers.

The solution values for the *cis* and *trans* proton couplings in the most stable conformer ($J_{b2,3}$ and $J_{b1,2}$) are practically independent of ΔS (see Table I). However, the solution values for these couplings in the least stable conformer ($J_{a2,3}$ and $J_{a1,2}$) are markedly dependent on ΔS. It is desired to use equation (6) to calculate $J_{a2,3}$ and $J_{a1,2}$ for each value of ΔS. The value of ΔS which gives the best agreement between $J_{a2,3}$ and $J_{a1,2}$, calculated using equation (6) and CONAL 2, is considered the best value.

Ia   Ib

In order to calculate $J_{a2,3}$ and $J_{a1,2}$ using equation (6), it is necessary to know values for θ, A, and X. We assumed $X = -0.02$, the value found for

Table I   Solution Parameters of Equation (1) Using Temperature-Dependences of Chemical Shifts and Coupling Constants and Fixed ΔS^a

Solution	ΔS (fixed, cal/mol-degree)	ΔH (cal/mole)	ν_{a1}	ν_{b1}	ν_{a2}	ν_{b2}	ν_{a3}	ν_{b3}	$\delta_{a1,3}$	$\delta_{b1,3}$	$\delta_{a1,2}$	$\delta_{b1,2}$	$J_{a2,3}$	$J_{b2,3}$	$J_{a1,2}$	$J_{b1,2}$
(a)	-0.5	-1843											5.67	11.68	4.91	3.48
	0.0	-1796											3.18	11.69	5.51	3.48
	0.5	-1755											1.40	11.70	5.93	3.48
		$(95)^b$											(0.79)	(0.03)	(0.22)	(0.01)
(b)	0.0	-1699	104.67	97.81	124.54	75.72	99.67	49.33								
	0.5	-1655	106.07	97.81	134.48	75.68	109.91	49.29								
		$(97)^b$	(1.42)	(0.07)	(5.39)	(0.23)	(5.55)	(0.24)								
(c)	-0.5	-1751	103.58	97.82	116.76	75.77	91.65	49.39					5.05	11.72	5.06	3.48
	0.0	-1700	104.68	97.82	124.59	75.73	99.72	49.33					3.77	11.73	5.36	3.47
	0.5	-1657	106.08	97.81	134.54	75.68	109.97	49.29					2.16	11.73	5.75	3.47
		$(74)^b$	(1.09)	(0.06)	(4.14)	(0.18)	(4.26)	(0.18)					(1.14)	(0.06)	(0.95)	(0.05)
(d)	-0.5	-1660							14.15	48.63	-11.04	22.23				
	0.0	-1607							7.73	48.68	-17.24	22.28				
	0.5	-1561							-0.42	48.73	-25.10	22.33				
		$(66)^b$							(2.26)	(0.15)	(2.33)	(0.16)				
(e)	-0.5	-1664							14.08	48.62	-11.11	22.24	5.44	11.75	4.97	3.47
	0.0	-1611							7.64	48.67	-17.32	22.28	4.27	11.76	5.25	3.46
	0.5	-1565							-0.54	48.72	-25.21	22.32	2.77	11.77	5.60	3.46
		$(46)^b$							(1.65)	(0.11)	(1.59)	(0.11)	(0.49)	(0.04)	(0.40)	(0.03)

^{a} All nmr parameters are given in cps. The chemical shift downfield from TMS of the nth proton in conformer a is $\nu_{a,n}$. The chemical shift between protons n and m in conformer a is $\delta_{a,n,m}$. Spectra were determined at 60 Mcps.

^{b} All parenthetical values are probable errors in the solution values of the parameters when ΔS was fixed. They are comparable to the probable error obtained when $\Delta S = 0.5$ eu. at 0.0 and -0.5 eu.

Table II   Comparison of $J_{a2,3}$ and $J_{a1,2}$ Calculated Using Equation (6) and CONAL 2

ΔS (cal/mole-degree)	$J_{b2,3}/J_{b1,2}$	ϕ_b (°)	A (cps)	$J_{a2,3}/J_{a1,2}$	ϕ_a (°)	Equation (6)		CONAL 2	
						$J_{a2,3}$	$J_{a1,2}$	$J_{a2,3}$	$J_{a1,2}$
−0.5[a]	3.37	56	11.6	1.000	59	3.0	3.0	5.05	5.06
0.0[a]	3.38	56	11.6	0.703	56	2.4	3.5	3.77	5.36
0.5[a]	3.38	56	11.6	0.376	51	1.6	4.4	2.16	5.75
1.0[a]	3.38	56	11.6	0.014	34	0.1	7.8	0.09	6.24
−0.5[b]	3.36	56	11.6	0.88	58	2.7	3.1	5.67	4.91
0.0[b]	3.36	56	11.6	0.58	54	2.1	3.8	3.18	5.51
0.5[b]	3.36	56	11.6	0.24	48	1.2	5.0	1.40	5.93

[a] Solution (c) of Table I.   [b] Solution (a) of Table I.

cyclohexane (*10*). For each fixed value of ΔS, θ_b was calculated from equation (7),

$$\frac{J_{b2,3}}{J_{b1,2}} = \frac{\cos^2(\omega + \theta_b) + X\cos(\omega + \theta_b)}{\cos^2 \theta_b + X\cos \theta_b} \quad (7)$$

which derives from equation (6), using the CONAL 2 solution values of $J_{b2,3}$ and $J_{b1,2}$ and $\omega = 118°$ (*10*). Then, a value of A was calculated using equation (6), $J_{b1,2}$ and θ_b. After estimating θ_a from equation (8)

$$\frac{J_{a2,3}}{J_{a1,2}} = \frac{\cos^2(\omega - \theta_a) + X\cos(\omega - \theta_a)}{\cos^2 \theta_a + X\cos \theta_a} \quad (8)$$

and the CONAL 2 solution values of $J_{a2,3}/J_{a1,2}$, $J_{a1,2}$ and $J_{a2,3}$ were calculated from equation (6) using the value of A determined before. The results of these calculations, some of which are summarized in Table II, show that the best agreement between values of $J_{a2,3}$ and $J_{a1,2}$, calculated using equation (6), and those computed using CONAL 2 is obtained when ΔS is about 0.5 eu.

We wondered whether or not low populations of the boat conformation of I (see Ic and Id) might influence significantly the solution values of ΔH given in Table I. Consequently, synthetic proton couplings were calculated for a three-conformer problem having $\Delta H = 5600$ cal/mole and $\Delta S = 1.38$ and 5 eu for equation (9) and $\Delta H = -1796$ cal/mole and $\Delta S = 0$ eu for equation (5). The average *cis* and *trans* proton couplings in Ic and Id were taken to be 4 and 11 cps, respectively, and solution (a) couplings in Table I ($\Delta S = 0$) were taken for the couplings in Ia and Ib. Using the synthetic proton couplings in CONAL 2, solution values of ΔH were obtained that fell within 30 cal/mole of that deriving from synthetic data which neglected the boat conformation; i.e., the two-conformer problem. It would appear from these results that the temperature-dependences of the nmr parameters of I can be used to solve the two-conformer problem (equation (5)) reliably.

$$(9)$$

APPENDIX : CONAL 2

The Newton-Raphson iterative least-squares treatment of CONAL 2 is in principle identical to the method described by Castellano and Bothner-By (3) for nmr spectral analysis. If the calculated experimental parameters, P_r^{calc}, are assumed to be linear functions of the unknown parameters, U_s, small changes in the unknown parameters, $\varDelta U_s$, result in linear changes in the calculated parameters, $\varDelta P_r^{\text{calc}}$. Thus,

$$\varDelta P_r^{\text{calc}} \simeq \frac{\partial P_r^{\text{calc}}}{\partial U_s} \varDelta U_s \tag{10}$$

where P_r^{calc} is equated to P_{ij}^{calc}, as given in equation (3), r refers to any combination of i and j, and U_s refers to any particular unknown. When $\varDelta P_r^{\text{calc}}$ is equated to the difference between the observed and calculated parameters, i.e., $\varDelta P_r^{\text{calc}} = P_r^{\circ} - P_r^{\text{calc}}$, $\varDelta U_s$ becomes the correction to the unknown, U_s, which when applied to the unknown would cause P_r° to equal P_r^{calc} if equation (10) was exact. In general, equation (10) is approximate so that $\varDelta U_s$ is an approximate correction. In addition,

$$\varDelta U_s = {}^1U_s - {}^0U_s$$

where ${}^0U_s$ is the value of the sth unknown in a given iteration and ${}^1U_s$ is the corresponding value in the next iteration. One would like to determine corrections in all the unknowns which, when applied to the unknowns, minimize ϕ in equation (2). Therefore, equation (10) is summed over all unknowns, giving

$$\varDelta P_r^{\text{calc}} = \sum_{s=1}^{n} \frac{\partial P_r^{\text{calc}}}{\partial U_s} \varDelta U_s \qquad \text{for } r = 1, 2, 3, \ldots, m$$

where n is the number of unknowns and m is the total number of observed parameters, P_{ij}°. The above equations of condition can be written in matrix notation as

$$\mathbf{D}\vec{\mathbf{U}} = \vec{\mathbf{P}}$$

where $\mathbf{D}$ is the $m \times n$ matrix of partial derivatives with elements $\alpha_{rs} = \partial P_r / \partial U_s$ obtained by differentiating equation (3), $\vec{\mathbf{U}}$ is the n-dimensional vector of corrections to the unknowns and $\vec{\mathbf{P}}$ is the m-dimensional vector of the residuals. Standard least-squares procedure (12) for minimizing ϕ is to form the normal equations

$$\mathbf{D}^{\text{T}}\mathbf{D}\vec{\mathbf{U}} = \mathbf{D}^{\text{T}}\vec{\mathbf{P}}$$

which are solved to yield the corrections to the unknowns, $\vec{U}$. Thus,

$$\vec{U} = (\mathbf{D}^T \mathbf{D})^{-1} \mathbf{D}^T \vec{P}$$

The corrections are applied to the unknowns and the iterative loop is repeated until all the corrections are less than 0.001 units.

Estimates of the errors in the calculated unknowns are made by estimating errors in the corrections to the unknowns when $\vec{U} = 0$, using the method of Castellano and Bothner-By (3). They use a standard statistical method where the standard deviation of the sth unknown is given by (13)

$$\sigma_s = (M_s \vec{P} \cdot \vec{P}/\det(\mathbf{D}^T \mathbf{D})(m - n))^{1/2} \tag{11}$$

where M_s is the minor of the coefficient $(\mathbf{D}^T \mathbf{D})_{ss}$ and $\vec{P} \cdot \vec{P}$ is the sum of the squares of the residuals. It is convenient to diagonalize the normal matrix, $\mathbf{D}^T \mathbf{D}$, before calculating the uncertainties (3). Equation (11) becomes

$$\sigma_q = (\vec{P} \cdot \vec{P}/d_{qq}(m - n))^{1/2}$$

where σ_q is the standard deviation of the qth linear combination of unknowns and d_{qq} is the qth diagonal element of the diagonalized matrix.

ACKNOWLEDGMENTS

We appreciate support of this work by the National Science Foundation (Grants GP-8315 and 5806).

REFERENCES

1. W.W. Wood, W. Fickett and J.G. Kirkwood, *J. Chem. Phys.* **20**, 561 (1952).
2. (a) H. Joshua, R. Gans and K. Mislow, *J. Am. Chem. Soc.* **90**, 4884 (1968) and references cited therein; (b) A.A. Bothner-By and D.F. Koster, *J. Am. Chem. Soc.* **90**, 2351 (1968); F. Heatley and G. Allen, *Mol. Phys.* **16**, 77 (1969).
3. S.M. Castellano and A.A. Bothner-By, *in* "Computer Programs for Chemistry," D.F. Detar, ed., Vol. I, pp. 10–39, Benjamin, New York, 1968.
4. H.S. Gutowsky, G.G. Belford and P.W. McMahon, *J. Chem. Phys.* **36**, 3353 (1962); R.S. Newmark and C.H. Sederholm, *J. Chem. Phys.* **39**, 3131 (1963); G. Govil and H.J. Bernstein, *J. Chem. Phys.* **47**, 2818 (1967); J. Jonas and H.S. Gutowsky, *J. Chem. Phys.* **42**, 140 (1965); H.S. Gutowsky, J. Jonas, F. Chen and R. Meinzer, *J. Chem. Phys.* **42**, 2625 (1965.
5. R.J. Abraham, L. Cavilli and K.G.R. Rachler, *Mol. Phys.* **11**, 471 (1966); R.J. Abraham and M.A. Cooper, *J. Chem. Soc.* (*B*) **1967**, 202.
6. N.L. Allinger, J.A. Hirsch, M.A. Miller, I.J. Tyminaki and F.A. Van Catledge, *J. Am. Chem. Soc.* **90**, 1199 (1968).
7. J.A. Hirsch, *Top. Stereochem.* **1**, 199 (1967); E.L. Eliel and T.J. Brett, *J. Am. Chem. Soc.* **87**, 5039 (1965).

8. K. Mislow, R. Graeve, A.J. Gordon and G.H. Wahl, Jr., *J. Am. Chem. Soc.* **86**, 1733 (1964); J.L. Coke and M.C. Mourning, *J. Org. Chem.* **32**, 4063 (1967).

9. E.W. Garbisch, Jr., *J. Chem. Ed.* **45**, 402 (1968).

10. E.W. Garbisch, Jr. and M.G. Griffith, *J. Am. Chem. Soc.* **90**, 6543 (1968).

11. M. Barfield and D.M. Grant, "Advances in Magnetic Resonance," Vol. 1, pp. 149–195, Academic Press, New York, 1965.

12. E.T. Whittaker and J. Robinson, "The Calculus of Observations," Blackie, London, 1929.

13. Reference 12, p. 246.

Conformations of Membrane-Active Cyclodepsipeptides

V.T. Ivanov and Yu.A. Ovchinnikov
Institute for Chemistry of Natural Products
USSR Academy of Sciences
Moscow, USSR

The use of various compounds capable of specifically modifying given properties of membranes (such as their permeability, excitability and the activity of their enzymatic constituents) has become one of the principal approaches to the physicochemical basis of membrane functioning. Among such membrane-affecting compounds, a special place is assumed by depsipeptides, especially valinomycin and enniatin groups (Fig. 1), which are antibiotics capable of inducing alkali metal permeability in various artificial and biological membranes (see reference 1 and references therein). As the work progressed it became obvious that the biological activity of the title compounds is intimately associated with their ability to form stable complexes with alkali cations; antimicrobial activity and stability constants of Na^+ and K^+ complexes are given for valinomycin (a), enniatin B (b) and some of their synthetic analogs (c–p) in Tables I and II. In turn, the essential role played by conformational properties of cyclodepsipeptides in manifesting the antimicrobial activity has been demonstrated (2). We therefore undertook a study of the conformational states of the cyclodepsipeptides and their complexes in solution on the example of valinomycin and enniatin B with the aid of various physicochemical methods (3, 4).

The ORD curves of valinomycin in different solvents differ significantly from each other, indicating the existence of various conformers in equilibrium, the latter being shifted with change in polarity of the medium (Fig. 2). Further information on the nature of the equilibrium was obtained from the nmr spectra of valinomycin in CCl_4 and $(CD_3)_2SO$ solutions (Fig. 3 (5, 6))*. These

---

* The nmr spectra were obtained on a JNM-4H-100 spectrometer operating at 100 Mc/sec. Assignment of $N_{(1)}H$ and $N_{(7)}H$ signals was made from the spectrum of valinomycin with one $N_{(7)}H$ group changed to $^{15}N_{(7)}H$ (see Fig. 3). The labeled sample of valinomycin was obtained by total synthesis starting from ^{15}N-L-valine.

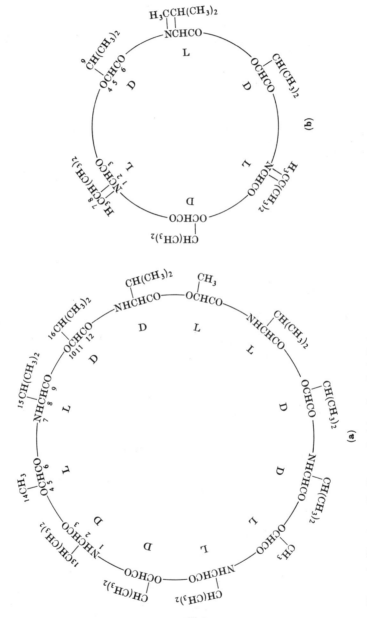

Figure 1 Structure of valinomycin (a) and enniatin B (b).

Table I Antimicrobial Activity and Complexation with K^+ of Valinomycin and Its Analogs

	Antimicrobial activity (minimal growth-inhibiting concentration, γ/ml)				Complexation[a] (stability constant, $K \cdot 10^{-5}$ liter/mole, EtOH, 25°C)
Compound	Staphylococcus aureus UV-3	Sarcina lutea	Mycobacterium phlei	Candida albicans	
a ⎡(D-Val-L-Lac-L-Val-D-HyIv)₃⎤ (valinomycin)	0.8	1.5	0.3	0.8	6.8
c D-Val→D-Leu; L-Val→L-Leu	8	22	1	8	6.0
d L-Lac→L-Ala	2	2	2	2	2.2
e L-Val→L-HyIv		Inactive	Inactive		0.0025
f L-Val→L-MeVal		Inactive	Inactive		0
g 3 L-Val→3 D-Val		Inactive	Inactive		0
h 3 D-HyIv→3 L-HyIv		Inactive	Inactive		0
i ⎡(L-Val-D-Lac-D-Val-L-HyIv)₃⎤ (enantio-valinomycin)	0.8	1.5	0.3	0.8	6.8
j ⎡(D-Val-L-Lac-L-Val-D-HyIv)₄⎤		Inactive			0.0001

[a] Compounds a and c–j do not form stable complexes with Na^+.

Table II   Antimicrobial Activity and Complexation of Enniatin B and Its Analogs

Compound	Antimicrobial activity (minimal growth-inhibiting concentration, γ/ml)				Complexation (stability constant, $K \cdot 10^{-3}$ liter/mole, EtOH, 25°C)	
	Staphylococcus aureus 209 P	Mycobacterium phlei	Sarcina lutea	Candida albicans	Na⁺	K⁺
b [L-MeVal-D-HyIv]₃ (enniatin B)	18	9–12	18	9–12	2.6	6.5
k L-MeVal→D-MeVal		Inactive			0.7	1.3
l D-HyIv→L-HyIv		Inactive			> 0.1	0.6
m [L-Val-D-HyIv]₃ (Tri-N-desmethyl)enniatin B		Inactive			2.5	2.6
n [L-MeLeu-D-HyIv]₃ (enniatin C)		Inactive			2.5	5.5
o [L-MeVal-D-HyIv]₄	> 50	18–25	3–4.5	> 50	1.0	2.8
p [D-MeVal-L-HyIv]₃ (enantio-enniatin B)	18	9–12	18	9–12	2.6	6.5

spectra clearly show the presence of two doublets corresponding to two different types of NH signals ($N_{(1)}H$ and $N_{(7)}H$). In CCl_4 these signals are relatively close to each other (δ 7.90 and 7.76 ppm, respectively) whereas in $(CD_3)_2SO$ the $N_{(7)}$ signal undergoes a low field shift of 0.6 ppm, $N_{(1)}H$ remaining practically unmoved. This served as a basis for the proposal that in CCl_4 the preferred conformation of valinomycin has six intramolecular H bonds formed by all

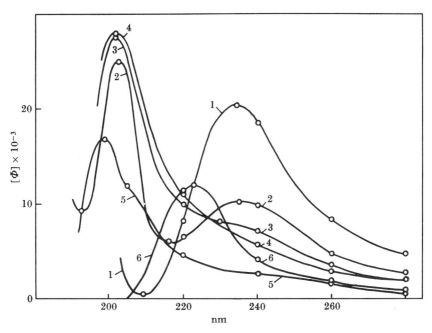

Figure 2 ORD curves of valinomycin and its K^+ complex (1:heptane–dioxane (10:1), 2:heptane–ethanol (3:1), 3:ethanol, 4:acetonitril, 5:trifluoroethanol–water (1:2), 6:ethanol + KBr).

NH groups, while in $(CD_3)_2SO$ the $N_{(7)}H$ protons form H bonds with the solvent molecules. This proposal is in accord with infrared (ir) spectral data of valinomycin in CCl_4 solution* which display strong bands at 3307 cm^{-1} due to intramolecularly H-bonded NH and much weaker bands at 3395 cm^{-1} due to free NH groups (Fig. 4). In the CO stretching region there is a symmetric band at 1757 cm^{-1} due to non-hydrogen-bonded ester carbonyls. Amide carbonyls display a band at 1661 cm^{-1} with an inflection at 1678 cm^{-1} and a

* Infrared spectra of valinomycin in CCl_4 are concentration-independent at least in the range $4 \cdot 10^{-2} – 4 \cdot 10^{-4}$ mole/liter, showing the absence of intermolecular association, also demonstrated by thermoelectrical weight measurements in various solvents over a wide range of concentrations.

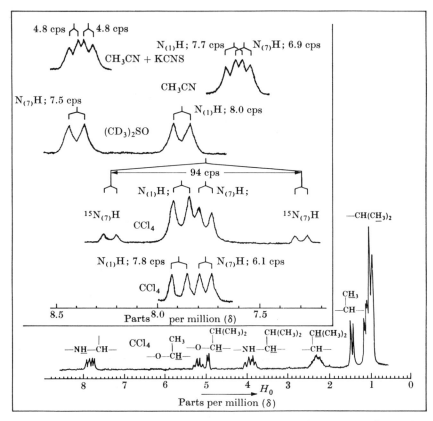

Figure 3 Nuclear magnetic resonance spectra of valinomycin and its K⁺ complex.

band at 1540 cm^{-1} corresponding to amide I and amide II bands. It is note-worthy that the presence of free and H-bonded amide groups are observed to about the same extent in the NH stretching frequency and amide I regions. All this bears witness to the fact that in low-polar media, valinomycin is in

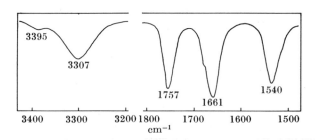

Figure 4 Infrared spectrum of valinomycin in CCl$_4$.

an equilibrium mixture of two forms (Fig. 5), of which one (A), with all amide groups forming six intramolecular H bonds, is clearly predominant, and the other (B), with only three such bonds increases in relative importance with increase in polarity of solvent.

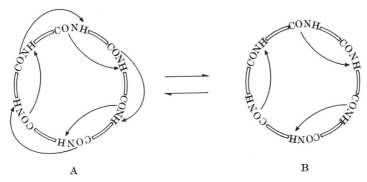

A   B

Figure 5   A ⇌ B equilibrium of valinomycin.

It can be readily shown that for *trans* configuration of the amide and ester groups in valinomycin,* the A conformation in which each carbonyl is involved in H bonding with the NH of the neighboring amide group in the "direction of acylation" is the only way by which all six amide groupings can participate in mutual H bonding. This is a quite rigid framework consisting of six condensed ten-membered rings formed by the H bonds, the whole resembling a "bracelet" ~8 Å in diameter and 4 Å high. However, the experimental data described above fit both the conformation A_1 and its inside-out counterpart A_2 (Fig. 6), the latter differing from the former in the chirality of the ring system and in orientation of the side chains.

The choice beween the A_1 and A_2 "bracelet" conformation of valinomycin was made on the basis of nmr data. It can be seen from molecular models that H–N–C^α and N–C^α–H dihedral angle of the NH–C^αH fragments in the A_2 conformation can vary within the limits 80–130°, which, according to our data (8), should correspond to a spin-coupling constant $^3J_{HN-CH} = 0.5$–5 cps. However, the experimental $^3J_{HN-CH}$ coupling constants are within the limits of 6.7–9.1 cps (taking into account the +0.6 cps correction for electronegativity). This not only excludes the A_2 conformation, but shows that one group of HN–C^αH fragments in the A_1 conformation ($^3J_{N_{(7)}H-C_{(8)}H} = 6.7$ cps in CCl_4) is gauche oriented, while the other ($^3J_{N_{(1)}H-C_{(2)}H} = 8.4$ cps in CCl_4) is *cis* oriented. This made possible determination of all six ester and amide carbonyls as three

Trans configuration of amide bonds follows from the intensive amide II band in ir spectra (Fig. 4); *cis* ester bonds are known to be present only in lactons with less than eleven atoms in the ring (7).

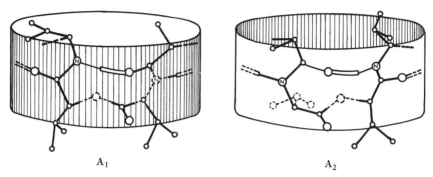

Figure 6 Schematic representation of forms A_1 and A_2 of valinomycin.

within and three without the ring. Finally, the value of the $^3J_{C^\alpha H-C^\beta H}$ constant showed the protons in the $C^\alpha H-C^\beta H$ fragments to be *trans* oriented in the amino acid and gauche oriented in the hydroxy acid residues.

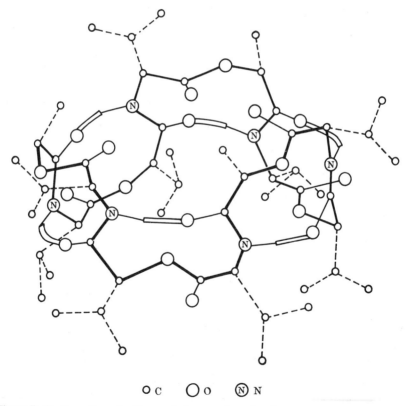

$\bigcirc$ C $\bigcirc$ O $\textcircled{N}$ N

Figure 7 Conformation of valinomycin in nonpolar solvents.

All these data served as evidence that the bracelet conformation presented in Fig. 7 is the one preferred in nonpolar media. The dipole moment calculated for this conformation is 2.5 ± 1.5 D, which is in good agreement with the experimental value of 3.5 ± 0.1 D (in CCl_4), especially if one takes into account the presence of a certain amount of the less symmetric form B.

Analogous physicochemical studies made it possible to determine the conformation of the K^+ complex in solution. The ir spectrum of this complex in $CHCl_3$ (Fig. 8) displays no free NH stretching band and the ester carbonyl frequency is shifted to the longer wavelengths by 16 cm^{-1} with simultaneous narrowing of the band (as compared with valinomycin). This indicated retention of the hydrogen-bond-stabilized framework in the K^+ complex of

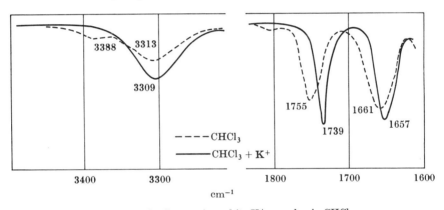

Figure 8   Infrared spectra of valinomycin and its K^+ complex in $CHCl_3$.

valinomycin, wherein all the ester carbonyls become involved in ion–dipole interaction with K^+. Complexation is accompanied by conformational reforming of the molecule in which the three ester carbonyls are now also oriented within the ring to form a hexadentate system of oxygen atoms about the cation. The ORD curve of the K^+ complex of valinomycin differs sharply from the ORD curves of both the A and B forms (Fig. 2). Moreover, the nmr spectra of the complex in all solvents investigated showed almost identical spin–spin coupling constants equal to 5.4 cps for all six HN–CH fragments (see Fig. 3), a sign of the gauche orientation of these protons in all cases. These results are in complete agreement with the rigid symmetric conformation of the K^+ complex of valinomycin presented in Fig. 9, a characteristic feature of which is effective screening by the hydrophobic peripheral side chains of the K^+ atom and the system of hydrogen bonds from solvent action.*

A similar approach was used in the investigation of conformational states of enniatin B (Fig. 1), the most popular member of the enniatin series, and its

* Very recently, this structure was confirmed by x-ray analysis (9).

K⁺ complex. Unlike valinomycin, enniatin B has all amide bonds N-methyl-
ated, which excludes the possibility of hydrogen bond formation and restricts
considerably the amount of information obtained from nmr and ir measurements. At the same time, enniatin B has twice smaller ring size than valinomycin, which enabled the theoretical conformational analysis to be carried
out in the way described below.

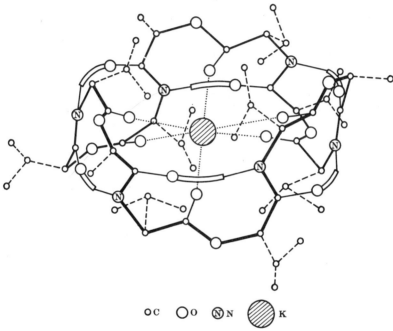

oc   Oo   Ⓝn   ⊘k

Figure 9   Conformation of the K⁺ complex of valinomycin.

The conformational maps (*10*) were computed for

$$\text{MeCO–N(Me)–(L)CH(CHMe}_2\text{)–COOMe (I)}$$

and

$$\text{MeCO–O–(D)CH(CHMe}_2\text{)–CONMe}_2 \text{ (II)}$$

by a method which has been described earlier (*11*), (I) and (II) modeling the
amino and hydroxy acid fragments of enniatin B. Using the minimization
procedure by variation of the angles $\phi_{C\alpha-N}$, $\phi_{C\alpha-O}$, $\psi_{C\alpha-C'}$, $\chi_{C\alpha-C\beta}$ and ν_{N-Me},
the favorable conformations of the molecules (I) and (II) have been calculated.
As follows from the results (Fig. 10) there are four iso-energetic minima
(k, l, m and n) on the potential surface of (I), whereas favorable conformations
(p and q) of (II) separated by a barrier of only 0.3 kcal/mole form one low-
energy area with $\phi_{(250-320)°}$, $\psi_{(50-100)°}$. The further analysis shows that

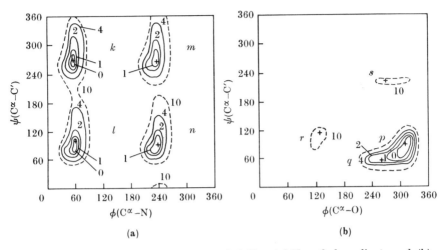

Figure 10 Conformational maps of (a) methyl-N-acetyl-N-methyl-L-valinate and (b) O-acetyl-D-α-hydroxyisovaleryl-N-dimethyl amide.

provided D-α-hydroxyisovaleryl residues of enniatin B have $\mathrm{II}_{p,\,q}$ conformation, there exist only two conformations of enniatin B, called N and P, with the orientations of N-methylvaline residues corresponding to potential minima on the conformational map of (I) (Fig. 11; ϕ and ψ values are approximate). Thus, the theoretical analysis strongly suggests that, in solution, enniatin B can assume two principal structures, not differing considerably in their energies.

Indeed, ORD studies of enniatin B in several solvents (Fig. 12a) indicate an existence of conformational equilibrium, its point depending upon the solvent. One conformer with a strong negative Cotton effect at 230 nm is predominant

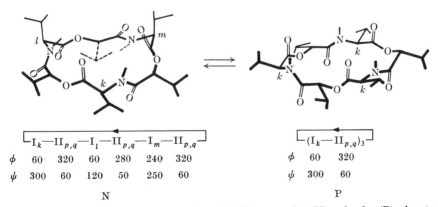

Figure 11 Preferred conformations of enniatin B in nonpolar (N) and polar (P) solvents.

in nonpolar solvents (heptane); as the polarity is gradually increased, the equilibrium is shifted until a new conformer becomes predominant with weak positive (~240 nm) and strong negative (~200 nm) Cotton effects. It is noteworthy that the conformation of the K^+ complex of enniatin B is of the same type as that of the noncomplexed antibiotic in polar solvents (trifluoroethanol), as follows from the similarity of corresponding ORD curves (Fig. 12a).

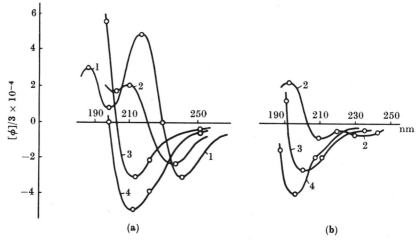

Figure 12 ORD curves of (a) enniatin B, (b) (tri-N-desmethyl)enniatin B, and their K^+ complexes (1:heptane, 2:ethanol, 3:trifluoroethanol, 4:ethanol + KCl).

Considerable information on the structure of "nonpolar" conformation was obtained from the nmr spectra. On cooling a solution of enniatin B in CS_2 or (2:1) CS_2–$CD_3C_6D_5$ mixture, the N-methyl signal splits into three singlets of equal intensity; similarly, the N-methylvalyl α-proton doublets split into three equal doublets, the middle ones overlapping with the α-proton doublets of the hydroxy acid (Fig. 13). From this it follows that although enniatin B is built up of three structurally identical subunits, all three N-methylvaline fragments in the above solvents have different spatial structure due to differences in rotation about N–C^α and C^α–C' single bonds (i.e., in the ϕ and ψ coordinates of the corresponding conformational maps). Taking into account the ϕ and ψ values of form N (Fig. 11), it was natural to assume that such conformation of enniatin B is dominating in nonpolar media; an excellent agreement of the calculated (3.5 ± 0.5 D) and measured in CCl_4 (3.35 ± 0.1 D) dipole moments served as a final proof for this structure. Large spin–spin coupling constants of the C^α–C^β protons (8.7–9.9 cps in nonpolar solvents, such as CCl_4, C_6H_6 and CS_2) indicated their preferable *trans* orientation. This compact structure has neither a central cavity nor symmetry elements; the *pseudo*-axial orientation of the three adjacent isopropyl groups (fragment

I_1–$II_{p,q}$–I_m) and *pseudo*-equatorial orientation of three other isopropyl groups (fragment $II_{p,q}$–I_k–$II_{p,q}$) is noteworthy.

The complexing conformation of enniatin B should be expected to possess, similarly to valinomycin, a central cavity capable of accommodating alkali

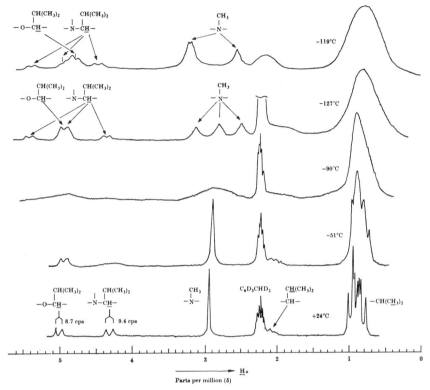

Figure 13   Nuclear magnetic resonance spectra of enniatin B in CS_2 at −119°C (upper spectrum) and in CS_2–$CD_3C_6D_5$ (2:1) at different temperatures (lower spectra).

cations, the six carbonyl groups pointing towards the interior of the molecule so as to provide efficient ion–dipole interaction with the central cation. These conditions are met by the structure P found in the course of theoretical analysis (Fig. 11). Therefore such conformation was assigned to the "complexing" conformation of enniatin B, this assumption having been confirmed by nmr studies of the complexes of (tri-*N*-desmethyl)enniatin B (compound m of Table II) with various univalent cations.* The use of this analog to determine

---

* A similar ("flat disc") conformation has been ascribed to the K^+ complex of enniatin B by Mueller and Rudin (*12*) on the basis of general considerations; recently Dobler *et al.* (*13*) were able to demonstrate that this conformation is also characteristic of the crystal (from x-ray analysis).

the conformation of enniatin B itself was justified by the fact that according to the ORD curves (Fig. 12), the two cyclodepsipeptides assume the same conformations on both dissolution in polar solvents and complexation.

In a comparison of the complexes of enniatin B and (tri-N-desmethyl)-enniatin B with cations of varying size, one would expect such complexes to differ in sizes of internal cavities formed by the oxygen atoms of the carbonyls participating in the ion–dipole interaction. Obviously, the effective size of the cavity for a given ring is determined by orientation of the carbonyl groups which in turn depends upon the cation radius. With small cations, the

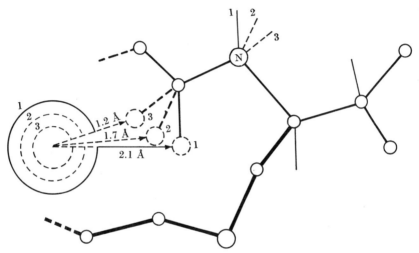

Figure 14   Effect of the size of the cation on $^3J_{\text{NH–CH}}$ coupling constants (cps) of (tri-N-desmethyl)enniatin B complexes (three different radii of the internal cavity are shown—1:9.0 cps, 2:6.6 cps, 3:3.8 cps).

closest distance between them would be with the carbonyls drawn into the center of the molecule. As larger cations are taken, all carbonyls are pushed more and more outwards in order to accommodate the cation so that the conformational changes ensuing could be likened to the opening of a flower bud. Analysis of molecular models shows that for the form P of (tri-N-desmethyl)enniatin B such change in orientation of the carbonyl groups inevitably leads to simultaneous rotation of the OC–NH plane (Fig. 14), the ultimate result of which would be increase in the dihedral angle between H–N–C$^\alpha$ and N–C$^\alpha$–H planes approximately from 130 to 160°, which change should be reflected in the nmr spectra by an increase of $^3J_{\text{HN–CH}}$ from 3.8 to 9.0 cps (8). Indeed, the nmr spectra of (tri-N-desmethyl)enniatin B with Li$^+$, Na$^+$, K$^+$ and Cs$^+$ displayed a monotonous increase in the $^3J_{\text{HN–CH}}$ constant (5.1, 6.5, 7.9 and 8.4 cps, correspondingly), showing that complexes of (tri-N-desmethyl)enniatin B and, consequently, enniatin B, are preferentially in P conformation. Further, from an analysis of the $^3J_{C^\alpha H-C^\beta H}$ coupling constants of the enniatin B·K$^+$ complex, it follows that the C$^\alpha$H–C$^\beta$H protons in the methylvaline residues are *trans* ($^3J_{C_{(2)}H-C_{(8)}H} = 9.8$ cps), whereas in the case of α-hydroxyisovaleryl residues, the populations of rotational isomers about C$^\alpha$–C$^\beta$ bonds are essentially averaged ($^3J_{C_{(5)}H-C_{(9)}H} = 6.5$ cps). The above data led to the conformation of the K$^+$ complex of enniatin B as depicted in Fig. 15.

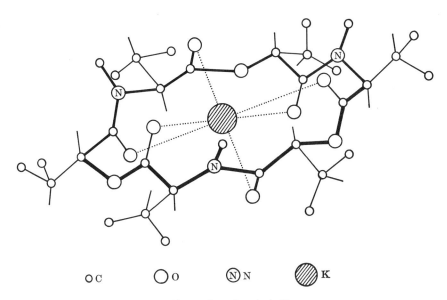

OC   O O   (N) N   K

Figure 15   Conformation of the K$^+$ complex of enniatin B.

The characteristic features of this conformation are the compact arrangement of the functional groups around the central cation, *pseudo*-equatorial orientation of all isopropyl groups and presence of symmetry axis; the exact shape of the "complexing" conformation is strongly dependent upon the radius of the cation.

Comparison of the two structures of valinomycin and enniatin B leads to an understanding of the differences in the efficiency and selectivity of their complexation (Tables I and II) and, consequently, of their regulation of membrane permeability. Although both valinomycin and enniatin B possess six carbonyl groups participating in the ion–dipole interaction with the central K$^+$, there is a considerable difference in the character of the ion–dipole interactions of these two compounds with K$^+$. As can be seen on Fig. 16, in the valinomycin·K$^+$ ester carbonyls are oriented almost directly towards the

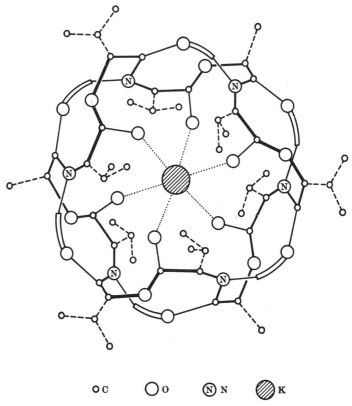

○ C   ◯ O   Ⓝ N   ⊘ K

Figure 16   Conformation of the K^+ complex of valinomycin (view along the symmetry axis).

cation, while in the K^+ complex of enniatin B "ion–dipole" bonds are almost perpendicular to C=O bonds. As the energy of ion–dipole interaction is known to be roughly proportional to $\cos\theta$

$$K^+\text{------}\overset{\theta\nearrow O^{\delta-}}{\underset{C^{\delta+}}{\big\|}}$$

the above-mentioned differences in the mutual orientation of the cation and carbonyl groupings result in different free energies of formation (−8.0 and −4.9 kcal/mole) (*1*) and stability constants ($6.8 \cdot 10^5$ and $3.7 \cdot 10^3$ liter/mole, Tables I and II) for the K^+ complexes of valinomycin and enniatin B.

    Contrary to the rigid, hydrogen-bond-stabilized conformation of the valinomycin molecule, enniatin B is characterized by a much higher conformational mobility. Therefore, in its complexing with Na^+ and K^+, the higher solvation energy of Na^+ ($89 - 72 = 17$ kcal/mole (*14*)) can be compensated by a more effective ion–dipole interaction brought about by turning the carbonyl groups so as to shorten oxygen-cation distances and straighten the angles θ (Fig. 14). All this should equalize the free energies of formation of enniatin B complexes with Na^+ and K^+, i.e., should lower the ionic selectivity. On the contrary, the rigid structure of valinomycin debars such conformational transition, which explains the exceptional K^+/Na^+ selectivity of this antibiotic.

The presented data explain many hitherto nonunderstood aspects of structure–activity relationship in depsipeptides. For instance, it became clear why individual amino and hydroxy acids in the molecule of valinomycin can be replaced by related compounds without loss of its complexing ability, but only up to the point when such replacement does not interfere with the bracelet conformation (compounds c, d and i of Table I). Such exchange is "forbidden" if it disrupts the system of intramolecular hydrogen bonds, as can be seen by the complete disappearance of the complexing and biological activity on replacement of an amino acid by a hydroxy acid or a N-methylamino acid residue (analogs e and f of Table I). It is interesting to note that in the cases when modification of valinomycin leads to noncomplexating analogs (compounds g, h and j of Table I) similar changes in the enniatin series are frequently followed by lowering, but not disappearing, complexing ability (compounds k, l and o of Table II). In other words, enniatin cyclodepsipeptides are characterized by comparatively low structural specificity of complexation as a result of their conformational lability, giving rise to the "induced fit" to the corresponding cation.

Investigation of macrocyclic depsipeptides enabled us to discover an earlier unknown mechanism, by which different peptide-related compounds devoid of ionizable functional groups are able to bind specifically alkali metal ions. This mechanism involves interaction of the ion with a spatially ordered system of amide or ester carbonyls so that the efficiency and selectivity of complexation depend to a great extent on the conformational parameters of the molecule. Since these compounds are structurally closely related to proteins, it is quite reasonable to expect that ion exchange regions of biological membranes responsible for their selective ionic permeability could make use of a similar principle for their functioning.

REFERENCES

1. M.M. Shemyakin, Yu.A. Ovchinnikov, V.T. Ivanov, V.K. Antonov, E.I. Vinogradova, A.M. Shkrob, G.G. Malenkov, A.V. Evstratov, I.A. Laine, E.I. Melnik and I.D. Ryabova, *J. Membrane Biol.* **1**, 402 (1969).
2. M.M. Shemyakin, Yu.A. Ovchinnikov, V.T. Ivanov, V.K. Antonov, A.M. Shkrob, I.I. Mikhaleva, A.V. Evstratov and G.G. Malenkov, *Biochem. Biophys. Res. Commun.* **29**, 834 (1967).

3.   V.T. Ivanov, I.A. Laine, N.D. Abdullaev, L.B. Senyavina, E.M. Popov, Yu.A. Ovchinnikov and M.M. Shemyakin, *Biochem. Biophys. Res. Commun.* **34**, 803 (1969).

4.   Yu.A. Ovchinnikov, V.T. Ivanov, A.V. Evstratov, V.F. Bystrov, N.D. Abdullaev, E.M. Popov, G.M. Lipkind, S.F. Arkhipova, E.S. Efremov and M.M. Shemyakin, *Biochem. Biophys. Res. Commun.* **37**, 668 (1969).

5.   D.H. Haynes, A. Kowalsky. B.C. Pressman, *J. Biol. Chem.* **244**, 502 (1969).

6.   M. Ohnishi, D.W. Urry, *Biochem. Biophys. Res. Commun.* **36**, 194 (1969).

7.   R. Huisgen and H. Ott, *Tetrahedron* **6**, 2531 (1959).

8.   V.F. Bystrov, S.L. Portnova, V.I. Tsetlin, V.T. Ivanov and Yu.A. Ovchinnikov, *Tetrahedron* **25**, 493 (1969).

9.   M. Pinkerton, L.K. Steinrauf and P. Dawkins, *Biochem. Biophys. Res. Commun.* **35**, 512 (1969).

10. J.T. Edsall, P.J. Flory, J.C. Kendrew, A.M. Liquori, G. Nemethy, G.N. Ramachandran and H.A. Scheraga, *Biopolymers* **4**, 121 (1966).

11. E.M. Popov, G.M. Lipkind, S.F. Arkhipova and V.T. Dashevsky, *Mol. Biol.* (*USSR*) **2**, 622 (1968).

12. P. Mueller and D.O. Rudin, *Biochem. Biophys. Res. Commun.* **26**, 398 (1967).

13. M. Dobler, J.D. Dunitz and F. Krajewsky, *J. Mol. Biol.* **42**, 603 (1969).

14. N.A. Izmailov, *Dokl. Akad. Nauk* (*USSR*) **149**, 1364 (1963).

Theoretical Conformational Analysis: ab initio SCF–LCAO–MO Studies of Conformations and Conformational Energy Barriers—Scope and Limitations

Jean-Marie Lehn
Institut de Chimie
Strasbourg, France

Ab initio, the procedure consisting in throwing together microphysical particles and laws of interaction and getting out complex molecular structures would be most satisfying. This is in principle possible since the Schrödinger equation (or better, the Dirac relativistic treatment) exists. However, reality efficiently tempers such idealistic rapture! Thus evolved the ab initio *self consistent field–linear combination of atomic orbitals–molecular orbitals* (SCF–LCAO–MO) theory (*1*) which may be considered as a close approach to the idealistic procedure (for a discussion of the present status of molecular calculations see references 2 and 3).

It is built on the following progressive steps (*2, 3*):

(1) Building up the molecular wave function from monoelectronic *molecular orbitals* obtained by linear combination of a basis set of atomic orbitals.

(2) Searching for self-consistency, i.e., for the best possible LCAO with respect to total energy.

(3) Improving over steps 1 and 2 by taking into account electron correlation, relativistic effects and molecular vibrations (vibronic structure).

Most of the present nonempirical molecular calculations, in particular those to be discussed here, do not go beyond point 2 and thus do not take into account electron correlation, relativistic effects and molecular vibrations. This raises a certain number of problems which will be considered below.

The purpose of this talk is to discuss briefly the present scope and limitations of using this theoretical model in *ab initio conformational analysis*, i.e., conformational analysis* starting from the laws and particles of matter (nuclei and electrons) and taking into account all interactions without using any empirical parameter.

Conformational problems arise only in relatively complex molecules so that such work has become practically feasible only since the advent of high-speed electronic computers.

We shall not go into theoretical details (1–3) here, remembering only that the important point for the present discussion is the nonempirical approach to the conformational problem. In fact, the process works forwards and backwards between methodology and conformations, the one permitting calculating the others, which in return serve as tests for the former.

We shall first discuss the scope and limitations and then the results of ab initio SCF–LCAO–MO conformational analysis as it has become feasible by marrying high-speed computers with efficient programs (see, for instance, references 4 and 5). Theoretical implications will be discussed in the context when necessary.

I. SCOPE AND LIMITATIONS

The first question to ask is when and why to use ab initio calculations in conformational analysis. One may roughly distinguish three types of approaches to conformational calculations:

(1) Phenomenological treatments, which try to simulate conformations using classical molecular mechanics and adjustable parameters (6).

(2) Semi-empirical quantum mechanical treatments at various levels of sophistication—extended Hückel theory (EHT) (7); semi-empirical SCF methods (Pariser-Parr-Pople (PPP) (8, 9); complete neglect of differential overlap (CNDO) (10); modified intermediate neglect of differential overlap (MINDO) (10b); electrostatic models (11).

(3) Strictly ab initio SCF–LCAO–MO calculations, introducing no empirical parameters, within the Hartree-Fock (HF) one-electron scheme, or including electron correlation (use of "geminals"; multiconfiguration technique) (2, 3).

Methods of type 1 lead to conformational information on the geometry, relative energies, etc., of various complex molecular conformations, and are *practically* very useful in many complicated cases, although they may be considered as quite artificial. However, the question here is getting quantitative information with a certain degree of confidence and not physical accuracy of the model.

---

* We use here *conformation* in a very general sense for designating the various spatial arrangements of the atoms forming a molecular system, which preserve the atomic connections in the molecular network. *Conformational analysis* is then the study of these conformations and of their interconversion paths.

Methods of type 3 are to be used when the physical *origin* of potential barriers, relative stabilities, etc., as well as the quantitative aspect are to be studied in terms of fundamental physical processes. Then, the physical soundness of the model is *the* problem.

Finally, methods of type 2 are somewhere between these two extremes, resembling type 1 in using empirical parameters and leading to relatively quick information (although often equivocal or erroneous (*12, 13*)), and resembling type 3 in being based on quantum mechanics and leading to deeper physical insight (*13*).

Thus ab initio methods are to be used when it is desired to describe conformational events in accurate microphysical terms, as they represent the physically realistic approach.

In a later step it is of course perfectly justified, and chemically useful, to extract from the ab initio description new conformational "concepts" or "effects" or to confirm former ones, which may be of great practical value. One may also try to calibrate phenomenological or semi-empirical methods or results with respect to ab initio ones.

Thus, ab initio conformational analysis is to convey physical bases to conformational thinking. This is so at least in principle; the problems it raises will now be discussed. There are two principal limitations in ab initio calculations of conformations:

(1) A theoretical limitation: all calculations have been done within the HF one-electron model which neglects electron correlation and relativistic effects.

(2) A practical limitation: the length of the computations, which leads to compromises between size of the molecules, basis set length and geometry optimization.

A. Electron Correlation—Relativistic Effects

In the HF one-electron model, a given electron interacts with the other electrons via their average field. While electrons with parallel spins (in different MO's) are kept apart by the Pauli principle, there is no electron–electron interaction term to prevent electrons with antiparallel spins, occupying the same region of space, from colliding with one another. The repulsive interactions between such electrons gives rise to an excess energy, called the correlation energy (*2, 3, 14*). The correlation energy generally amounts to 0.5% or less of the total energy.

For the lighter elements, the relativistic term is even smaller by about a factor of ten; however, it increases with the atomic number and becomes larger than the correlation term for silicon and higher elements (*3*). In view of the very small conformational energies (energy barriers and energy differences between conformers are in most cases below 50 kcal/mole), it is necessary that the correlation energy (which amounts to ~230 kcal/mole for NH_3) (15) and the relativistic correction (~15 kcal/mole for NH_3) be *very* nearly the same for the various conformations of a molecule.

Luckily this invariance requirement of the correlation energy, as well as of the relativistic effects, seems to be fulfilled. Justifications for this conclusion stem principally from the fact that calculations approaching the Hartree-Fock limit yield geometries in close agreement with the experimental ones and satisfactorily reproduce energy barriers (inversion barrier in ammonia (15); rotation barrier in ethane (16)) including some fine details in energy functions (for example, in the H_2O_2 internal rotation energy profile (17)) (Tables I–III).

Thus, although these justifications are a posteriori, it seems that one can rely with a satisfactory degree of confidence on the assumption of correlation energy invariance (or near invariance).* The same argument applies to the relativistic term.†

B. Choice of the Basis Set of Atomic Orbitals

Two principal problems arise with respect to the choice of the basis set of atomic orbitals which will be used to build up the molecular orbitals: its size and its "balance" (20).

A basis set of infinite size would lead to a Hartree-Fock wave function, and then any discrepancy with experiment would arise from correlation and relativistic effects.

In practice, if results are to be obtained whose quality will not depend on the type of molecule studied, it is highly desirable that the basis set be as large as possible. In terms of the gaussian type of basis functions (which will be considered here)‡ and from results obtained on atoms (21), it appears that in order to reach near Hartree-Fock wave functions, basis sets of $11s$ and $7p$ for C, N and O atoms and $6s$ for the H atom are required (basis set 11.7/6).

Furthermore, the addition of polarization functions (20) like $2p$ on H and $3d$ on C, N and O, which convey a greater flexibility to the basis set, is also required.

---

* In fact, even for chemical reactions taking place with conservation of the number of electron pairs, changes in correlation energy seem to be quite small (of the order of 6 kcal/mole) (18). However, only complete treatments can definitely settle this point.

† Another indication about the invariance of the relativistic term may be found in the relativistic treatment of the H_2 molecule by Kolos and Wolniewicz (19). The electronic kinetic energy and the relativistic energy show a similar dependence on H–H bond length. A change of ~600 kcal/mole in kinetic energy corresponds to a variation of only 0.01 kcal/mole in relativistic energy. Kinetic energy changes being in general much smaller than 600 kcal/mole (of the order of 100 kcal/mole or less) from one conformation of a molecule to another, the relativistic term may be considered as invariant.

‡ Several types of atomic basis functions have been used: Slater-type orbitals (STO), gaussian-type functions (GTF), gaussian-lobe functions (GLF).

It is often not practically feasible to use such large basis sets; wave functions further away from the HF limit are then obtained. As a consequence, the results are less reliable and the energy terms obtained may be much better for one molecule than for another; for instance, the barrier to internal rotation is less affected in H_3C-CH_3 than in $HO-OH$ by changes in the basis set (Table I).

Table I Barriers to Internal Rotation: ab initio SCF–LCAO–MO Results

Molecule	Barrier (kcal/mole)	Basis set[a]	Reference
H_3C-CH_3	3.3–3.5	STO (2.1/1)	50
	3.60	ext. GTF (10.6/4)	24
	3.62	ext. GTF (10.6/5)	24
	2.51	ext. GLF (10.5/4)	36
	2.88	v. lim GFT (5.2/2)	40
	3.45	lim. GFT (5.2/2 + p on H)	40
	3.05	lim. GTF (7.3/3)	31
	3.65	ext. GTF (11.7.1/6.1)	16
	3.07	ext. GTF (11.7.1/6.1) + g. opt.	16
	3.3	STO (2.1/1) + g. opt.	91
	3.31	STO (2.1/1)	116
	3.30	STO (2.1/1) + second-order perturbation	116
	2.928	experiment	51
$(H_3C-CH_3)^+$	2.5	ext. GTF (11.7.1/6.1) (by extrapolation)	16
H_3C-CH_2F	2.59	ext. GTF	38
	3.33	experiment	52
H_3C-CHF^+	0.62	lim. GTF (7.3/3) (–CHF$^+$ planar)	102
$H_2FC-CH_2^+$	10.53	lim. GTF (7.3/3) (–CH$_2^+$ planar)	102
$H_3C-CH_2^+$	0.0	ext. GTF (9.5/5) (–CH$_2^+$ planar)	53
	2.8	ext. GTF (9.5/5) (–CH$_2^+$ tetrahedral)	53
H_3C-NH_2	2.42	ext. GLF (10.5/4)	36
	2.02	v. lim GFT (5.2/2)	40
	1.98	experiment	54
H_3C-OH	1.35	ext. GLF (10.5/4)	36
	1.06	ext. GLF (11.6/4)	36
	1.59	v. lim. GFT (5.2/2)	40
	1.07	experiment	55
H_3C-SiH_3	1.44	ext. GTF (10.6.1(C)12.9.1(Si)/5.1)	16
	1.71	experiment	56
$H_3C-CHCH_2$	1.48	v. lim. GTF (5.2/3)	41
	1.98	experiment	57
	3.6	v. lim. GTF (4.3/2)	104
H_3C-CHO	0.8	?	quoted in 38

(continued)

Table I—*Continued*

Molecule	Barrier (kcal/mole)	Basis set[a]	Reference
$H_3C–CHO$	1.48	ext. GTF (9.5/4)	97
	1.16	experiment	98
$H_2N–CHO$	21.7	ext. GTF (11.7.1/6.1)	35
	20.2	experiment (dimethyl formamide)	59
$H_3C–NO$	1.88	ext. GTF (10.5/4)	60
	1.05	ext. GTF	95
	1.137	experiment	96
$H_3C–N(O)H$	1.07	ext. GTF (9.5/4)	97
	0.93		
$H(O)C–OH$	14.2	v. lim. GTF (5.2/2)	99
	9.46 (*cis–trans* energy difference)		
	10.9	experiment	100
	2.0 (*cis–trans* energy difference)	experiment	
$H_2N–NH_2$	11.5 (*cis*)	ext. GTF (9.3/3)	61
	4.7 (*trans*)		
	11.88 (*cis*)	ext. GLF (10.5/4)	37
	3.70 (*trans*)		
	11.05 (*cis*)	v. lim. GTF (5.2/2)	40
	6.21 (*trans*)		
	3.14	experiment assuming equal *cis* and *trans* barriers	62
$H_2N–OH$	11.95 (*trans* to *cis*)	ext. GLF (10.5/4)	37
	1.16 (*cis* to *trans*)		
	9.90 (*trans* to *cis*)	v. lim. GTF (5.2/2)	40
	2.53 (*cis* to *trans*)		
$HO–OH$	11.8 (*cis*)[b]	STO (2.1/1)	63
	2.2 (*trans*)[b]		
	13.62 (*cis*)	ext. GLF (10.5/4)	36
	13.27 (*cis*)	ext. GLF (11.6/4.1)	36
	0.1 (*trans*)		
	13.08 (*cis*)	STO (2.1/1)	64
	15.94 (*cis*)	v. lim. GTF (5.2/2)	40
	0.1 (*trans*)	lim. GTF (5.2/2.1)	40
	14.7 (*cis*)	ext. GTF (11.7.1/6.1)	17
	0.23 (*trans*)		
	10.9 (*cis*)	ext. GTF (11.7.1/6.1) + g. opt.	17
	0.6 (*trans*)		
	9.4 (*cis*)	STO (2.1/1) + g. opt.	91
	0 (*trans*)		
	7.0 (*cis*)	experiment	65
	1.1 (*trans*)		
$HS–SH$	7.4 (*cis*)	GTF (13.6/3)	93

Table I—*Continued*

Molecule	Barrier (kcal/mole)	Basis set[a]	Reference
HS–SH	1.9 (*trans*)		
	9.33 (*cis*)	GTF (12.9.1/5.1)	92
	5.99 (*trans*)		
	almost equal barriers	experiment	94
H_2P–PH_2	2.32 (*cis*)	lim. GTF (8.4/2)	105
	0.51 (*trans*)		
	2.57 (*cis*)	lim. GTF (7.3.1/2)	105
	0.49 (*trans*)		
	2.17 (*cis*)	lim. GTF (7.3/2)	105
	0.66 (*trans*)		
H_3N—BH_3	3.21	ext. GTF (9.5/4)	25
	3.29	ext. GTF (9.5.1/4)	25
	3.06	ext. GTF (11.7.1/6.1)	16
H_3C–$S(O)H$	2.68	v. lim. GTF (3.1/1; 5.2.1 for S)	66
H_2C^-–$S(O)H$	7.6	v. lim GTF (3.1/1; 5.2.1 for S)	66
$HC(-OH)_2^+$	*trans,trans–cis,trans* energy difference:		
	7.11	v. lim. GTF (5.2/2)	99
	1.5	v. lim. GTF (3.2/2)	45
	0.5	experiment (in solution)	101
	cis,cis–cis,trans energy difference:		
	5.67	v. lim. GTF (5.2/2)	99
	7.5	v. lim. GTF (3.2/2)	45
H_2C^-–SH	18.8	v. lim. GTF (3.1/1; 5.2.1 for S)	103a
	6.1		
H_2C^-–SO_2H	several barriers and energy differences	v. lim. GTF (3.1/1; 5.2.1 for S)	103b
H_2C^-–S–CH_2^-	several barriers and energy differences	v. lim. GTF (3.1/1; 5.2.1 for S)	103a
H_3C–CH_2CH_3	3.42	v. lim. GTF (5.2/2)	29
	3.48	lim. GTF (7.3/3)	29
	3.57	experiment	29
H_3C–$CH_2CH_2CH_3$	2.92	v. lim. GTF (5.2/2)	30
	2.94	lim. GTF (7.3/3)	30
	3.25–3.40	"experiment"	30, 67
H_3CH_2C–CH_2CH_3	3.54 (*trans → gauche*)	v. lim. GTF (5.2/2)	30
	8.82 (*gauche–cis*)	v. lim. GTF (5.2/2)	
	0.82 (*trans–gauche* energy difference)	v. lim. GTF (5.2/2)	
	3.62 (*trans → gauche*)	lim. GTF (7.3/3)	30
	6.83 (*gauche–cis*)	lim. GTF (7.3/3)	
	0.76 (*trans–gauche* energy difference)	lim. GTF (7.3/3)	

(*continued*)

Table I—*Continued*

Molecule	Barrier (kcal/mole)	Basis set[a]	Reference
$H_3CH_2C-CH_2CH_3$	3.6–3.9 (*trans →* *gauche*)	experiment	reference in 30
	5.3; 6.5–6.7 (*gauche–cis*)	experiment	
	0.77 (*trans–gauche* energy difference)	experiment	

[a] Abbreviations: ext., extended basis set; lim., limited basis set; v. lim., very limited basis set; STO, Slater type orbitals; GTF, gaussian type functions; GLF, gaussian lobe functions; g. opt., geometry optimization. Basis set composition is given in parentheses with the following convention: number of s functions·number of p functions·number of d functions on heavy atom/number of s functions·number of p functions on H: $(s.p.d/s.p)$. Several GTF's or GLF's are needed to represent one STO.

[b] These results may be in error; see reference 64.

When it becomes necessary to use limited basis sets, the results have first to be appreciated critically. However, satisfactory results may nevertheless be obtained in some cases, as will be seen below. In addition, *basis set optimization* may permit reducing its length.*

It often happens that extending the basis set greatly affects the total energy of systems, but affects much less relative energies like barriers or relative stabilities (*24, 25*.)

This is, however, not always the case; polarization functions may contribute more to one form than to the others, for instance, for symmetry reasons, and this may markedly affect energy differences. Thus the inclusion of d orbitals has a large effect (factor 2 or more) on the inversion barriers of CH_3^- (*26, 27*) and NH_3 (*15*) (in PH_3, however, the relative effect is much smaller (*28*); factor 1.2) (Table III). The presently available results seem to indicate that a large basis set containing only s and p functions leads to less satisfactory results than a smaller s, p set to which polarization functions have been added.

* The exponents of the basis functions are generally taken from optimized atomic (*21*) or small molecule calculations. In the case of a complete exponent optimization the major problem is the computer time required for performing it. It is often preferable to split an exponent and use two functions rather than optimize the exponent of a single function. Another problem arises from the fact that the GTF's are often grouped into a set of *contracted* gaussian functions (CGTF). It is then necessary to choose the *type* of contraction (for instance the GTF basis set 10.6/5 may be contracted to 5.3/3 or to 4.2/2, etc.) and the *coefficients* of the GTF's making up a given CGTF. Both are generally taken from atomic calculations (*21, 22*) or from model calculations on small molecules. The contraction does not in general seriously affect the quality of the calculation, if the type of contraction is carefully chosen (*23*).

The problem of building up a "well-balanced" (20) basis set is of course a consequence of the use of relatively small basis sets and disappears progressively as the set increases towards "saturation." It seems generally not recommended to introduce polarization functions on one atom and not on the others (for instance, on C and not on H or conversely). However, the final choice needs some practice and consideration of the computations reported in the literature.

For instance, satisfactory results have been obtained recently in a study of conformations and energy barriers in hydrocarbons using a very small basis set; this is likely to be due to the fact that the basis set had first been optimized on methane and that hydrocarbons seem to represent a particularly favorable case (29–31) (Tables I and II).

Table II   Conformational Energies in Cyclic Systems: ab initio SCF–LCAO–MO Results

Molecule	Conformation	Relative energy (kcal/mole)	Basis set[a]	Reference
Cyclobutane	Planar	0.00	STO (2.1/1)	106
	Planar	1.28	experiment	120
Cyclopentane[b]	Half-chair	0.00	v. lim. GTF (5.2/2)	31
	Envelope	0.24		
	Planar	8.07		
Cyclohexane	Chair	0.00[c]	v. lim. GTF (5.2/2)	31
	Twist-boat	6.05		
	Boat	7.19		
	Transition	11.22		
	Chair	0.00[d]	ext. GTF (9.5/5)	68
	Twist-boat	7.26		
	Twist-boat	4.8–5.5	experiment	69
	Transition	10.8	experiment	70

[a] Abbreviations: see footnote a in Table I.

[b] Ring puckering barrier and relative energy of planar form in tetrahydrofurane are respectively equal to 0.17 and 3.5 kcal/mole (71).

[c] Total energy $= -232.91$ au (1 au $= 627.71$ kcal/mole).

[d] Total energy $= -233.96$ au.

C. Geometry Optimization

A really complete ab initio calculation requires also that all geometrical parameters (bond lengths, bond angles) be optimized for all conformations. This of course becomes intractable for even quite simple molecules. If the calculations are of good quality, the geometry of conformational ground states*

* "Ground state" is used here in opposition to conformational "transition state," both states belonging to the *electronic* ground state of the molecule under consideration.

should be and is found to be (see, for instance, references 15–17 and 28) close to the experimental geometry, so that optimization may become of secondary importance and unnecessary. This is the case for PH_3, where a near Hartree-Fock calculation leads to the experimental bond length and bond angle (Table IV) (*28*).

Transition-state geometries should in principle be optimized in all cases, and even fine details in the potential functions may then be reproduced (*17*). However, in many cases, less accurate but still satisfactory results may be obtained when the same geometrical parameters (bond lengths, bond angles) are used for different conformations.*

D. Conclusions—Use of Models

The above brief discussion leads to the following conclusions based on the presently available computational results:

(1) Correlation energy and relativistic energy changes between conformations are of minor importance in ab initio conformational analysis.

(2) Large basis sets should be used if reliability and accuracy is expected (at least a Slater double ζ or a $(9s.5p/4s)$ type gaussian basis); the inclusion of polarization functions is recommended and may often be of much importance (especially for processes of inversion at atomic sites).

(3) Geometry optimization is generally not of prime importance unless very accurate energy values are desired, except in special cases (e.g., vicinal heteroatoms as in H_2O_2 (*17*)).

Good calculations are always time-consuming. But complex and interesting molecules should not just be discarded as intractable. A compromise has to be found which tries to take into account computer time and chemical interest of the problem. On the other hand, excellent agreement between experimental and calculated values may accidentally be obtained from rather poor calculations.

The specificity of ab initio SCF–LCAO–MO conformational analysis lies in the possibility of drawing a *physical picture* of the processes studied. Such a picture is expected to be less sensitive to the details of the calculations (basis set, geometry) than the quantitative results (see also Section IIC). In this respect, it may sometimes be preferable to use a *model system* rather than a real (i.e., synthetically available) but larger molecule, as a high-quality calculation on a model molecule may lead to more physical insight than an

* It is possible that a certain amount of the calculated geometrical changes may be due to the limitations of the computation. This is of course the case, when accurate experimental geometries are not reproduced.

average or mediocre calculation on a real but larger molecule. The problem is then the choice of the model, which relies more on chemical intuition than on brute computer force!

II. RESULTS

We shall successively discuss the following aspects of the conformational problem: conformational energies, geometries and physical origin of the conformational energy barriers within the ab initio HF–SCF–LCAO–MO framework. Each of these topics will be analyzed with respect to processes interconverting various conformations: internal rotations, inversions in cyclic molecules, inversions at specific sites (nitrogen, carbanion, oxonium, phosphorus).

Only the most significant or accurate calculations on a given system will be considered. In addition, as we have been primarily concerned with inversion processes at atomic sites in molecules (especially nitrogen inversion), this topic will receive comparatively more attention than internal rotations on which most of the ab initio conformational work has been done up to now.

The results for internal rotation, ring inversion, site inversion and geometry are listed in Tables I, II, III and IV, respectively.

Table III   Nitrogen, Carbanion, Oxonium, Phosphorus and Carbon Radical Inversion Barriers: ab initio SCF–LCAO–MO Results

Molecule	Barrier (kcal/mole)	Basis set[a]	Reference
NH_3	9.23	STO (2.1/1)	72
	91.64	Hartree-Fock AO's (2.1/1)	73
	1.69	21 STO's (one center) + g. opt.	74
	25.80	25 STO's (one center)	75
	11.53	STO (2.1/1)	76
	−2.14	STO (4.2/1)	76
	2.70	ext. STO (one center)	77
	1.69	ext. GTF (9.5/4)	43
	0.63	ext. GLF (10.5/5)	78
	0.80	ext. GTF (11.7/6) + $2p_z$ on H + g. opt.	15
	6.43	ext. GTF (11.7.1/6) + $2p_z$ on Ḣ	15
	7.41	ext. GTF (11.7.1/6) + $2p_z$ on H + g. opt.	15
	5.77	experiment	79
	5.08	ext. GTF (13.8.2/8.1)	108
	2.4	ext. STO (4.2/2)	109, 110

(*continued*)

Table III—*Continued*

Molecule	Barrier (kcal/mole)	Basis set[a]	Reference
NH_3	5.2	ext. STO (4.2/2) + CI	109, 110
H_2N–H′	Model calculations with different H–N–H angles and different charges on H′ nucleus	ext. GTF (10.5/5)	107
H_2N–F	20.3	ext. GTF (NH_2:9.5.1/4.1; F: 9.5)	111
	12.6	ext. GTF (9.5/4)	111
H_2N–CN	47.7	v. lim. GTF (5.1/1)	113
	Planar	ext. GTF (9.5/4)	111
	1.85	ext. GTF (9.5.1/4.1)	111
	1.33–2.03	experiment	114
H_2N–SiH_3	Planar N site	ext. GTF (SiH_3:12.9/5; NH_2:10.6/5)	112
	0.8	ext. GTF (SiH_3:12.9/5; NH_2: 10.6.1/5.1)	112
	0.65	ext. GTF (SiH_3:12.9.2/5; NH_2:10.6.1/5.1)	112
H_2N–CHO	1.4	ext. GTF (10.5/4)	80
	1.0	GTF (7.3/3) + $1s$ on each H of NH_2	81
	0.1	ext. GTF (11.7.1/5.1) + g. opt.	35
	1.1	experiment	82
Aziridine ⟨N—H⟩	15.5	v. lim. GTF (5.2/2)	quoted in 46
	18.3	ext. GTF (9.5/4)	32
	16–19	experiment (on derivatives)	quoted in 32
⟨N—H⟩	35.14	v. lim. GTF (5.2/2)	quoted in 83
Oxaziridine ⟨O,N—H⟩	32.4	ext. GTF (9.5/4)	32
	~32	experiment (on derivatives)	32, 84, 85
H_2C=N–H	27.9	ext. GTF (9.5/4)	33, 34
	26.2	ext. GTF (10.6/5)	33, 34
	25–27	experiment (on derivatives)	quoted in 33, 34
H–N=N–H	50.1	ext. GTF (9.5/4)	33, 39
H–N=C=N–H	8.4	ext. GTF (9.5/4)	33, 39
	6.7	experiment (on derivative)	85
CH_3^-	Planar	ext. STO (one center)	86
	1.22	ext. GTF (10.6/4)	26
	3.5	ext. GTF (10.6/8) + g. opt.	26

Table III—*Continued*

Molecule	Barrier (kcal/mole)	Basis set[a]	Reference
CH_3^-	7.0	ext. GTF (10.6.1/8) + g. opt.	26
	5.2	ext. GTF (10.6.2/5.1) + g. opt.	27
⟨cyclopropenyl⟩ C^-—H	20.85	v. lim. GTF (5.2/2)	46
⟨cyclopropyl⟩ C^-—H	52.3	v. lim. GTF (5.2/2)	83
$H_2C{=}C^-{-}H$	38.9	ext. GTF (10.6/5)	34
$H_2C{=}O^+{-}H$	17.2	v. lim. GTF (3.2/2)	45
$\cdot CH_3$	Planar	ext. GTF (10.6.2/5.1)	27
	Planar	GTF (5.2/2)	115
	Planar	minimal GTF	118
$H_2C{=}\dot{C}{-}H$	7.97	ext. GTF (10.6/5)	27
Cyclopropyl radical	3.8	ext. GTF (9.5/4)	97
$\cdot CH_2F$	0.6	GTF (5.2/2)	115
$\cdot CF_3$	27.4	GTF (5.2)	115
	21	ext. GTF (9.5)	117
BH_3^-	0.7	minimal GTF	118
NH_3^+	Planar	minimal GTF	118
SiH_3^-	39.6 (36.0)[b]	ext. GTF	119
PH_3	25.9	ext. STO (one center) + g. opt.	74
	30.9	ext. GTF (12.9/5)	28
	38.2	ext. GTF (12.9.2/5.1)	28
	37.2	ext. GTF (12.9.2/5.1) + g. opt.	28
	40.4 (30.9)[b]	ext. GTF	119
SH_3^+	30.0 (17.6)[b]	ext. GTF	119

[a] Abbreviations: see footnote *a* in Table I. Nearly all calculations have been performed with optimization of angles at the inverting site; some calculations also include bond length optimizations (noted g. opt.: geometry optimization).
[b] Barriers calculated with and without (value in parentheses) *d* polarization functions on the inverting site.

Table IV Conformational Geometries: ab initio SCF–LCAO–MO Results

Molecule	Bond lengths $r(A{-}B)$ (Å) and bond angles $A{-}B{-}C$ (°)		Conformation	Basis set[a]	Reference
$H_3C{-}CH_3$	$r(C{-}C)$:	1.551	Staggered	(11.7.1/6.1)	16
		1.570	Eclipsed	(11.7.1/6.1)	16
		1.543		experiment	75
	$\angle\,H{-}C{-}H$:	107.31	Staggered	(11.7.1/6.1)	16

(continued)

Table IV—*Continued*

Molecule	Bond lengths $r(A-B)$ (Å) and bond angles $A-B-C$ (°)		Conformation	Basis set[a]	Reference
H_3C-CH_3	$\angle$ H–C–H:	106.97	Eclipsed	(11.7.1/6.1)	16
		109.3		experiment	87
HO–OH	r(O–O):	1.500	*cis*	(11.7.1/6.1)	17
		1.475	*trans*	(11.7.1/6.1)	17
		1.411	*cis*	STO (2.1/1)	91
		1.408	*trans*	STO (2.1/1)	91
		1.475		experiment	88
	r(O–H):	0.993	*cis*	STO (2.1/1)	91
		0.987	*trans*	STO (2.1/1)	91
	$\angle$ O–O–H:	106.21	*cis*	(11.7.1/6.1)	17
		101	*trans*	(11.7.1/6.1)	17
		104.6	*cis*	STO (2.1/1)	91
		99.6	*trans*	STO (2.1/1)	91
		94.8		experiment	88
	Dihedral angle:	123		(11.7.1/6.1)	17
	(energy minimum)	111–120		experiment	88
H_2N-CHO	r(C=O):	1.211	Planar	(11.7.1/6.1)	35
		1.203	Perpendicular	(11.7.1/6.1)	35
		1.193		experiment	82
	r(C–N):	1.363	Planar	(11.7.1/6.1)	35
		1.418	Perpendicular	(11.7.1/6.1)	35
		1.376		experiment	82
NH_3	r(N–H):	1.020	Pyramidal	ext. STO (one center)	74
		1.012[b]	Pyramidal	(11.7.1/6) + $2p_z$ on H	15
		1.004	Planar	(11.7.1/6) + $2p_z$ on H	15
		1.00	Pyramidal	ext. GTF (13.8.2/8.1)	108
		0.984	Planar	ext. GTF (13.8.2/8.1)	108
		1.0116		experiment	89
	$\angle$ H–N–H:	108.90	Pyramidal	ext. STO (one center)	74
		106.2	Pyramidal	(11.7.1/6) + $2p_z$ on H	15
		107.2	Pyramidal	ext. GTF (13.8.2/8.1)	108
		106.70		experiment	89
CH_3^-	r(C–H):	2.090	Pyramidal	(10.6.1/8)	26
		2.038	Planar	(10.6.1/8)	26
	Out of plane angle:	23.5	Pyramidal	(10.6.1/8)	26
PH_3	r(P–H):	1.414	Pyramidal	est. STO (one center)	74
		1.442	Pyramidal	(12.9/5)	82
		1.421	Pyramidal	(12.9.2/5.1)	28
		1.402	Planar	(12.9/5)	28
		1.385	Planar	(12.9.2/5.1)	28
		1.421		experiment	90

Table IV—*Continued*

Molecule	Bond lengths $t(A-B)$(Å) and bond angles $A-B-C(°)$		Conformation	Basis set[a]	Refer-ence
PH$_3$	∠ H–P–H:	89.80	Pyramidal	ext. STO (one center)	74
		97.20	Pyramidal	(12.9/5)	28
		93.83	Pyramidal	(12.9.2/5.1)	28
		93.83		experiment	90

[a] Abbreviations: see footnote *a* in Table I.
[b] This value has been corrected using Table IX in reference 15.

A. Conformational Energies

Ab initio calculations lead to the total energies of the various conformations of a molecule, which may be either stable forms (valleys in the energy function) or transition states. Thus relative stabilities of conformational isomers and energy barriers to conformational interconversion processes are obtained. Except for some recent work, most calculations have been concerned with energy barriers.

Good agreement between calculated and experimental conformational energies for rotamers or invertomers may be fortuitous when small basis sets are used; oscillations may occur in the calculated energy differences as the basis set is progressively extended.

From Tables I and II it is seen that hydrocarbons represent a particularly favorable case, where calculations using small or large basis sets yield similar energies. This is clearly apparent in the case of ethane, for which a great variety of basis sets lead to similar barriers to internal rotation.

The same seems to be true for cyclic hydrocarbons (see cyclohexane in Table II). Thus it appears that nonempirical calculations may be performed even on relatively complex acyclic or cyclic systems, using small basis sets but leading nevertheless to satisfactory energy values.

Difficulties are encountered when the molecule contains heteroatoms bearing electronic "lone pairs," and especially when two such atoms are adjacent. In such cases it seems quite clear that accurate results will only be obtained with large, flexible basis sets including polarization functions.

The case of the barrier to rotation about the O–O bond in H$_2$O$_2$ is particularly enlightening in this respect (Table I). Clearly, more numerical experimentation is necessary on such systems where electron distributions are likely to be strongly affected by geometrical changes.

Satisfactory energies are obtained for rotations about C–X bonds with relatively small basis sets.

Inversion processes represent a difficult case, as the changes imposed upon the system on going to the transition state are comparatively larger than in the case of rotation barriers.

The difficulties encountered with nitrogen inversion are exemplified by the great variety of values obtained for the inversion barrier of ammonia (Table III).

Recent calculations on NH_3 *(15)* as well as on PH_3 *(28)* clearly show that in order to obtain satisfactory results, polarization functions (especially *d* functions) have to be included in the basis set in these two systems on both the central atom X and on H; *d* functions on X favor the pyramidal state and *p* functions on H seem to balance to some extent the effect of the *d* functions. The use of an unbalanced basis set for NH_3 (only *p* on H) leads to poor results (Table III). An unpolarized set may lead to better results than a polarized but unbalanced set.

Of course high barriers to inversion are comparatively easier to reproduce than low ones; the absolute discrepancies are of the same order of magnitude in both cases, but the relative ones are much more important for NH_3 than for PH_3. Thus one might expect that the addition of polarization functions has less pronounced effect on high barriers.

In the case of larger systems like aziridine *(32)*, oxaziridine *(32)*, methyl-enimine *(33, 34)*, satisfactory agreement between the calculated inversion barriers and available experimental data is obtained; again the barriers are quite high in these cases. The experimental trends are well reproduced (Table III), but cancellation of errors cannot be excluded. The rigidity of the molecular framework holding the nitrogen site (as compared to acyclic amines) is probably of importance in obtaining satisfactory results. The inversion barriers calculated for the isoelectronic carbanion and oxonium sites should be subject to similar remarks, but only a few unrelated data are presently available, except for CH_3^- *(26, 27)* which indeed seems to present the problems found in NH_3.

B. Conformational Geometries

In most calculations, only the geometrical parameter leading to the process studied (rotation angle; pyramid height or angle for inversions) was being varied. Of course, in principle complete geometry optimization should be performed for each conformation. This has been done in a few cases (Table IV). An interesting feature of performing such optimization is the possibility of obtaining the geometries of conformational transition states.*

For hydrocarbons, geometry optimization does not seem to be of prime importance, as seen from a recent calculation on ethane without and with

---

* Although in many cases the *rate* of conformational interconversion may be markedly increased by tunneling through the barrier, the height of the barrier is nevertheless determined by the energy of a geometrically defined transition state.

optimization of the C–C bond length and of the H–C–H bond angle (16) (Table IV); however, in the last case an improved value of the barrier is obtained, the C–C bond becoming longer and the H–C–H angle smaller in the eclipsed form. However, in H_2O_2 the small *trans* barrier in the internal rotation energy curve is only obtained (together with a much improved value for the *cis* barrier) when geometry optimization is performed, even when extended polarized basis sets are used (17). Not unexpectedly, a longer O–O bond and an opening of the O–O–H angle are found for the *cis* form (transition state) (Table IV).

Inversion barriers in XH_3 compounds are also affected by $r(X–H)$ optimization. Shorter X–H bonds are found for the planar transition states. It is also interesting to note that, in agreement with simple reasoning, the C=O and C–N bond lengths in H–CO–NH_2 (35) are, respectively, shorter and longer (Table IV) in the perpendicular transition state for rotation about the C–N bond than in the planar ground state.

C. Physical Picture of Conformations and Conformational Energy Barriers

Perhaps the most important contribution of ab initio calculations to the understanding of conformational phenomena is their potential ability to provide a physically sound picture of the origin of conformational properties. Thus one may hope that a physical picture of the origin of conformational energy barriers may be found within the HF–SCF–LCAO–MO framework, and recent results indicate that this is indeed the case for internal rotation (36–38) and inversion (28, 32–34) processes. Such a qualitative picture is also expected to be less sensitive to computational parameters (basis set, geometries) than the quantitative value of the barrier. However, detailed testing of this expectation is still needed.

We shall consider successively the following two aspects of the problem: energetical origin of barriers and electronic changes accompanying conformational processes. Of course, the calculations also provide a detailed description of the electronic structure of each individual conformer; but the present discussion will be centered on the changes brought about by the interconversion of conformers.

1. *Energy-component analysis of conformational energy barriers.* The total energy of a molecule E_{total} is a sum of four terms: the nuclear repulsion potential V_{nn}, the nuclear–electron attractions V_{ne}, the electron–electron repulsions V_{ee} and the electronic kinetic energy T:

$$E_{total} = V_{nn} + V_{ne} + V_{ee} + T$$

The variation of each of these terms may be studied in the course of conformational change.

However, at present, it is only possible to describe qualitatively the origin of the conformational energy barriers in terms of energy components, if one uses a decomposition of the total energy into components whose *comparative* changes (phase, amplitude) on going from one conformation to another are not much affected by improving the wave function, by scaling the energy terms to force agreement with the virial theorem* and by small geometrical perturbations (*36, 38*).

Several two-component decompositions of the total energy may be considered:

(1) V_{ne} (attractive) and $V_{nn} + V_{ee} + T$ (repulsive);
(2) $V_{nn} + V_{ne} + T$ (attractive) and V_{ee} (repulsive);
(3) $V_{ne} + V_{ee} + T$ (electronic; attractive) and V_{nn} (nuclear; repulsive);
(4) $V_{ne} + V_{ee} + V_{nn}$ (potential; attractive) and T (kinetic; repulsive).

The energy barriers may then be described in terms of the variations ΔX and ΔY of the two components X and Y in a given decomposition. The stabilities of these ΔX and ΔY values under the three perturbations mentioned above are the same neither for all decompositions, nor for different types of barriers except in the case of decomposition (1), whose ΔX and ΔY display remarkable stability. In the following discussion we shall therefore make use of decomposition (1), which has been introduced recently by Allen (*38*) for studying the barriers to internal rotation.†

The comparative changes ΔA in the *attractive* term A (V_{ne}) and ΔR in the *repulsive* term $R(V_{nn} + V_{ee} + T)$, between the ground and the transition states are much less stable under the three perturbations mentioned above than the change in total energy itself (i.e., the energy barrier) (*38, 40*); they are, however, stable enough to provide a qualitative physical picture of the energy barriers.

Depending on which variation, ΔA or ΔR, is the larger, the barrier may be considered as being, respectively, of attractive or of repulsive origin. Furthermore, the variations of the A and R terms are of opposite phase as one goes over the barrier: when one component increases, the other one decreases.

* The presently available approximate wave functions do not satisfy the virial theorem $E_{total} = -T = V/2$ (where $V = V_{nn} + V_{ne} + V_{ee}$). One may force agreement with this theorem by introducing a scale factor $\eta = -V/2T$ for each conformation (*36, 37, 39, 40*). This is much less satisfying than using improved wave functions which ultimately will obey the virial theorem.

† In many cases it would also be of interest to consider decompositions (2), (3) and (4); for instance, (2) permits comparing changes in bielectronic repulsions in related systems. In the present state of the problem (especially because of the lack of very good wave functions) the energy-component analysis of the barrier origin cannot be as complete as one might wish; one is limited to the use of components which are not too sensitive to the computational limitations. Of course, the ultimate would be a direct comparison of the four terms making up E_{total}.

Two types of conformational energy barriers may thus be distinguished:

(1) The barriers due to the fact that R decreases less than A increases (i.e., becomes less negative, less attractive) on going towards the top of the barrier; conversely, one may also consider that the ground state is more stabilized by A than destabilized by R with respect to the transition state; these barriers may be called *"attractive barriers,"* as they are governed by the A term

$$(|\Delta A| > |\Delta R|)$$

(2) The barriers due to the reverse effect: R increases more than A decreases (i.e., becomes more negative, more attractive) on going from ground state to transition state; they may be called *"repulsive barriers,"* as they are dominated by the R term $(|\Delta R| > |\Delta A|)$.

This may be summarized in the following scheme:

Component \\ Barrier type	A (V_{ne})	R $(V_{nn} + V_{ee} + T)$
Attractive	↗	↘
Repulsive	↘	↗

(The variations indicated correspond to going from the most stable form towards the top of the barrier.) It is the subtle balance between the variations of these two A and R components which determines the type and the height of the barrier.

Barriers to internal rotation. The physical picture of these barriers has been studied by Allen *et al.* (*36, 37*).

Repulsive barriers are found in ethane (*36*), methanol (*36*), methylamine (*36*) ethyl fluoride (*38*), propane (*41*), hydrazine (*37, 38*) (both *cis* and *trans* barriers).

The barrier to internal rotation in acetaldehyde (*38*) and the larger barrier in the hydroxylamine (*37*) and in the hydrogen peroxide (*36*) are of the *attractive* type, while the smaller barrier of hydroxylamine (*37, 38*) is found to be repulsive-dominant.*

* The nature of the weak barrier in H_2O_2 is not yet known although it is very probably repulsive-dominant (*38*).

Both extended (*36, 37*) and limited basis set (*40*) calculations on these compounds lead to the same physical image of the barriers, thus supporting the idea that this image is relatively insensitive to the length of the basis set used.

A detailed analysis of the rotation process in the various compounds leads to a description of the barriers in terms of interactions between bonds, lone pairs and protons (*36–38*).

It will now be of interest to await the detailed analysis of the ethane and hydrogen peroxide results of Veillard (*16, 17*), to see if this physical picture is conserved on introduction of polarization functions in the basis set and on geometry optimization.

A recent *bond-function* analysis of the rotational barrier in ethane (*42*) also points to a *repulsive* origin of this barrier. The dominant term is found to be the repulsive overlap interaction between bond orbitals.

Inversions at atomic sites. We have been mainly concerned with studying the physical origin of inversion barriers, especially *nitrogen inversion* barriers (*32–34, 39, 43*).

The inversion barrier in NH_3 is of the *repulsive* type; this seems to be true both when the basis set does not contain (*43*) and when it includes (*44*) polarization functions.

The barriers to nitrogen inversion in aziridine (*32*), oxaziridine (*32*), methylenimine (*33, 34*), carbodiimide (*33, 39*) and diimide (*33, 39*) are found to be of the *attractive* type. In comparison to NH_3, the attractive forces favor the ground state in these compounds. The nuclear–electron attraction V_{ne} stabilizes preferentially "compact" conformation where the nuclei and electrons are closer together; thus the high barrier (Table III) found in the "strained" aziridine arises from the fact that the small CNC angle renders the molecule more compact and stabilizes the pyramidal form, where the N–H bond comes closer to the rest of the molecule. This is even more so in oxaziridine, which is slightly more pyramidal (*32*), and where (in the present energy analysis, scheme (1)) the higher barrier, as compared to aziridine, is due to increased V_{ne} stabilization of the ground state with respect to the transition state.*

The *carbanion inversion* barriers in CH_3^- (*27*) and in the vinyl anion (*34*) are both of the *attractive* type.

The *phosphorus inversion* barrier in PH_3 (*28*) is of the repulsive type. In addition, an important point is that this result holds as well in the absence and in the presence of polarization functions but is dependent on geometry optimization.

---

* Energy decomposition (2) adds another aspect to this comparison by showing that the bielectronic repulsions are comparatively more important in the transition state of oxaziridine than in that of aziridine, with respect to the corresponding ground states.

The *oxonium inversion* barrier in $CH_2{=}\overset{+}{O}H$ seems to be of the repulsive type (*34, 45*).

In the isoelectronic series $CH_2{=}XH$, with $X = C^-$, N or O^+, the barrier to inversion at X increases in the order $O^+ < N < C^-$. As the physical pictures of C^- and N inversions are similar, one would expect that a pyramidal C^- site should have a higher inversion barrier than a pyramidal N site, and by extension, that a pyramidal O^+ site should have a lower barrier than a N site. This seems indeed to be the case in systems of type

where a larger barrier has been calculated for $X = C^-$ (*46*) than for $X = N$ (*32, 46*) (Table III), and where a lower barrier is experimentally obtained for a derivative of $X = O^+$ (~10 kcal/mole) (*47*).

More computational results are needed for guaranteeing the generality of these results (especially for carbanion, oxonium and phosphorus inversions) and for ascertaining the effects of substituents. The individual energy terms may also be considered separately; for instance, in the case of nitrogen inversion in aziridine (*32*) the V_{ne} attractions stabilize and the terms V_{nn}, V_{ee} and T destabilize the ground state with respect to the transition state.

The nitrogen inversion process in aziridine, oxaziridine and methylenimine has also been studied in terms of *localized molecular orbitals* (*48*). The main conclusion (*48*) is that no specific bond (nor lone pair) interaction is responsible for the inversion barrier, but all pairwise interactions contribute to it. In other words, analyzing the inversion process in terms of localized MO's clearly shows that the barrier origin itself is delocalized over the whole molecule (for a more detailed discussion see references 48 and 127).

Pseudorotation. The pseudorotation process in PH_5 has been studied recently (*44*) (basis net 12.7.1/5). A barrier of 3.9 kcal/mole has been obtained for the interconversion of two trigonal bipyramidal forms through a square pyramidal transition state. The barrier is of the repulsive type.

2. *Electronic distribution rearrangement on conformational interconversion.* The rearrangements of the electronic distribution on changing molecular conformation may be studied by performing a population analysis in terms of atomic and overlap populations (*49*). Overlap populations measure bond strengths as well as attractions or repulsions between atoms (*49*).

One should not attach too much weight to the quantitative values of the populations, as they are known to depend markedly on the basis set used, but general trends in population changes during conformational interconversion may present interesting features, which may be useful in characterizing conformations and conformational processes.

Internal rotation. On rotating from the staggered to the eclipsed form, the C and H atoms in ethane, respectively, gain and lose electron population, the C–H bonds becoming slightly more polar *(40, 50)*.

One also finds that in CH_3-XH_n compounds, where X is a heteroatom bearing electronic lone pairs, the atomic population of a given hydrogen of the CH_3 group loses electron density to a given lone pair as it rotates closer to it *(40)*.

The same holds for XH_n-YH_m compounds. Overlap populations have been used to describe the rotation of the methyl group in propene *(41)*; the total overlap population is larger for the more stable conformation. The same holds for ethane *(50)*.

Inversion processes. The general trends in electron distribution changes during nitrogen inversion in ammonia *(15, 43)*, aziridine *(32)*, oxaziridine *(32)* and methylenimine *(33, 34)* may serve to characterize the process. One observes the following changes:

(1) The nitrogen lone pair MO becomes a pure $2p$ orbital in the transition state.

(2) A $N(2s) \rightarrow N(2p)$ electron transfer into the lone pair p orbital is observed; a large transfer corresponds to a high inversion barrier, although there is no direct correlation between electron transfer and barrier height, except in isoelectronic series like compounds $CH_2=XH$, with $X = C^-$, N or O^+, where the barrier height and the changes in $X(2s)$ and $X(2p)$ populations increase in the sense $O^+ < N < C^-$ *(34)*.

(3) The total atomic population on nitrogen and on the hydrogen linked to it, respectively, increase and decrease during inversion.

(4) The bond overlap populations increase during inversion, i.e., the bonds become stronger.

(5) The total overlap population is larger in the transition state than in the most stable pyramidal form. This result contrasts with what is observed for internal rotation in propene *(41)* and in ethane *(50)* (see above).

Similar changes are observed for carbanion inversion in CH_3^- *(26)* and in the vinylic anion *(34)*, for oxonium inversion in $CH_2=\overset{+}{O}H$ *(45)* (but the O–H overlap population decreases in the transition state) and for phosphorus inversion in PH_3 *(28)*. In carbodiimide an electron transfer from the inverting nitrogen to the vicinal C=N double bond is observed; such a transfer is in line with the familiar picture of a lone pair–double bond conjugation, which also leads to a lowering of the inversion barrier (as compared to methylenimine) *(33)*.

III. CONCLUSION

Several conclusions may be drawn from the results and the discussion presented above:

(1) Ab initio SCF–LCAO–MO calculations of good quality are able to reproduce conformational energies (relative stabilities of conformers; energy barriers to the interconversion of conformers) and geometries.

(2) Ab initio conformational analysis may be performed on relatively complex systems and leads to answers in satisfactory agreement with experiment.

(3) A physical picture of the origin of the energy barriers hindering conformational interconversion processes may be found within the HF–SCF–LCAO–MO framework. The energy changes may be analyzed in terms of a set of components of the total energy (attractive and repulsive components, localized bond energy terms, etc.).

(4) Such energy-component analysis may provide a physical picture of both barriers to *internal rotation* and barriers to *site inversion* (nitrogen, carbanion, oxonium, phosphorus, etc.); it very probably also holds for ring inversion and pseudorotation processes. It may also permit a qualitative understanding of the relative barrier heights and structural effects.

It should be stressed that such a qualitative scheme represents only a first step towards a fundamental understanding of molecular conformations. More computations and numerical experimentations are needed especially for establishing the criteria for performing meaningful calculations (basis sets, geometries, models, etc.).

Also, the fundamental questions of the correlation and of the relativistic energy changes from one conformer to another remain still to be answered, although there are some reasons (which are probably more than just wishful thinking!) for considering these changes as being very small. Going beyond the Hartree-Fock scheme is more than ever an urgent necessity.

Thus, in the coming years, ab initio conformational analysis will probably develop along two main lines: one going deeper into the physical nature of the problem by including the correlation energy and ultimately the relativistic contribution; the other line branching out into chemistry by treating more and more complex systems within the Hartree-Fock or similar schemes, and trying to extract more and more detailed physical pictures from the computational results.

The present status of the question is certainly not the ultimate one! But on going from empirical to nonempirical treatments of the conformational problem a decisive step forward is being made in the understanding of the physical nature of molecular conformations and of conformational processes.

Note added in proof. In view of the rapid expansion of the field discussed above, some comments will be added here.

(1) Partial inclusion of electron correlation through a second-order perturbation method leaves the barrier to rotation in ethane virtually unchanged (*116*).

(2) In recent calculations (*109, 110*) of the NH_3 inversion barrier, the improvement brought about by configuration interaction is probably an artefact of the calculation, as the basis set used does not include d polarization functions on the inverting nitrogen site. Such functions are of fundamental importance in the computation of inversion barriers (especially low ones), as shown by recent results on NH_3 (*107*), H_2N–F (*111*), H_2N–CN (*111*) and H_2N–SiH_3 (*112*) (see Table III).

(3) Inner shell correlation effects are negligible (*125*).

(4) It has recently been found that in some cases (H_2O_2, H_2S_2 (*17*); PH_3 (*39*)), the *attractive* dominant–*repulsive* dominant analysis of energy barriers (Section IIC above) (*38*) is not stable on basis set extension or geometry optimization. For a discussion of the physical picture or "origin" of energy barriers, see also reference 127.

(5) The above discussion did not include rotation about double bonds. Such data may be found in the literature (see for instance, ethylene (*121*), allene (*121–123*).

(6) Polarization functions and balance of basis sets of gaussian functions have been discussed recently (*124*).

(7) Inversion processes at atomic sites have recently been the subject of three general discussions (*126–128*).

REFERENCES

1. C.C.J. Roothaan, *Rev. Mod. Phys.* **23**, 69 (1951).
2. E. Clementi, *J. Chem. Phys.* **46**, 3842 (1967).
3. E. Clementi, *Chem. Rev.* **68**, 341 (1968).
4. I.G. Csizmadia, M.C. Harrison, J.W. Moskowitz, S. Seung, B.T. Sutcliffe and M.P. Barnett, "POLYATOM," Program 47.1, Quantum Chemistry Program Exchange, Indiana University, Bloomington, Ind.
5. D.R. Davies and E. Clementi, "IBMOL," Program 92, Quantum Chemistry Program Exchange, Indiana University, Bloomington, Ind; A. Veillard, "IBMOL Version 4," Special IBM Tech. Rpt., San Jose, Ca., 1968.
6. J.E. Williams, P.J. Strang and P. von R. Schleyer, *Ann. Rev. Phys. Chem.* **19**, 531 (1968) and references therein.
7. R. Hoffmann, *J. Chem. Phys.* **39**, 1397 (1963); **40**, 2474, 2480, 2745 (1964).
8. R. Pariser and R.G. Parr, *J. Chem. Phys.* **21**, 466 (1953); **23**, 711 (1955).
9. J.A. Pople, *Proc. Phys. Soc. (London)* **A 68**, 81 (1954).
10. J.A. Pople, D.P. Santry and G.A. Segal, *J. Chem. Phys.* **43**, S 129 (1965); J.A. Pople and G.A. Segal, *J. Chem. Phys.* **43**, S 136 (1965); (a) J.A. Pople, *J. Chem. Phys.* **43**, S 229 (1965); (b) N.C. Baird and M.J.S. Dewar, *J. Chem. Phys.* **50**, 1262 (1969); N.C. Baird, M.J.S. Dewar and R. Sustman, *J. Chem. Phys.* **50**, 1275 (1969).

11. J.P. Lowe and R.G. Parr, *J. Chem. Phys.* **44**, 3001 (1966); **45**, 3059 (1966).

12. W.C. Herndon, J. Feuer and L.H. Hall, *Tetrahedron Letters* **1968**, 2625; J.R. De La Vega, Y. Fang and E.F. Hayes, *Intern. J. Quantum Chem.* **III S**, 113 (1969).

13. M.S. Gordon, *J. Am. Chem. Soc.* **91**, 3122 (1969).

14. C. Hollister and O. Sinanoglu, *J. Am. Chem. Soc.* **88**, 13 (1966).

15. R.G. Body, D.S. McClure and E. Clementi, *J. Chem. Phys.* **49**, 4916 (1968).

16. A. Veillard, *Chem. Phys. Letters* **3**, 128 (1969).

17. A. Veillard, *Chem. Phys. Letters* **4**, 51 (1969); *Theoret. Chim. Acta* **18**, 21 (1970).

18. L.C. Snyder and H. Basch, *J. Am. Chem. Soc.* **91**, 2189 (1969).

19. W. Kolos and L. Wolniewicz, *J. Chem. Phys.* **41**, 3663 (1964).

20. R.S. Mulliken, *J. Chem. Phys.* **36**, 3428 (1962).

21. S. Huzinaga, *J. Chem. Phys.* **42**, 1293 (1965).

22. A. Veillard, *Theor. Chim. Acta* **12**, 405 (1968).

23. C. Salez and A. Veillard, *Theor. Chim. Acta* **11**, 441 (1968).

24. E. Clementi and D.R. Davies, *J. Chem. Phys.* **45**, 2593 (1966).

25. M.-C. Moireau and A. Veillard, *Theor. Chim. Acta* **11**, 344 (1968).

26. R.E. Kari and I.G. Csizmadia, *J. Chem. Phys.* **46**, 4585 (1967); **50**, 1443 (1969).

27. P. Millié and G. Berthier, *Intern. J. Quantum Chem.* **II S**, 67 (1968).

28. J.M. Lehn and B. Munsch, *Chem. Commun.* **1969**, 1327.

29. J.R. Hoyland, *J. Chem. Phys.* **49**, 1908 (1968).

30. J.R. Hoyland, *J. Chem. Phys.* **49**, 2563 (1968).

31. J.R. Hoyland, *J. Chem. Phys.* **50**, 2775 (1969).

32. J.M. Lehn, B. Munsch, P. Millié and A. Veillard, *Theor. Chim. Acta* **13**, 313 (1969).

33. J.M. Lehn and B. Munsch, *Theor. Chim. Acta* **12**, 91 (1968).

34. J.M. Lehn, B. Munsch and P. Millié, *Theor. Chim. Acta* **16**, 351 (1970).

35. B. Bak, private communication.

36. W.H. Fink and L.C. Allen, *J. Chim. Phys.* **46**, 2261, 2276 (1967).

37. W.H. Fink, D.C. Pan and L.C. Allen, *J. Chem. Phys.* **47**, 895 (1967).

38. L.C. Allen, *Chem. Phys. Letters* **2**, 597 (1968).

39. J.M. Lehn and B. Munsch, unpublished results.

40. I. Pedersen and K. Morokuma, *J. Chem. Phys.* **46**, 3941 (1967).

41. M.L. Unland, J.R. Van Wazer and J.H. Letcher, *J. Am. Chem. Soc.* **91**, 1045 (1969)

42. O.J. Sovers, C.W. Kern, R.M. Pitzer and M. Karplus, *J. Chem. Phys.* **49**, 2592 (1968).

43. A. Veillard, J.M. Lehn and B. Munsch, *Theor. Chim. Acta* **9**, 275 (1968).

44. A. Rauk, L.C. Allen and K. Mislow, to be published.

45. P. Ros, *J. Chem. Phys.* **49**, 4902 (1968).

46. D.T. Clark and D.R. Armstrong, *Chem. Commun.* **1969**, 850.

47. J.B. Lambert and D.H. Johnson, *J. Am. Chem. Soc.* **90**, 1349 (1968).

48. B. Lévy, P. Millié, J.M. Lehn and B. Munsch, *Theor. Chim. Acta* **18**, 143 (1970).

49. R.S. Mulliken, *J. Chem. Phys.* **23**, 1833, 1841, 2338, 2343 (1955).

50. R.M. Pitzer and W.N. Lipscomb, *J. Chem. Phys.* **39**, 1995 (1963); R.M. Pitzer, *J. Chem. Phys.* **41**, 2216 (1964); **47**, 965 (1967).

51. S. Weiss and G.F. Leroi, *J. Chem. Phys.* **48**, 962 (1968).

52. P.H. Verdier and E.B. Wilson, *J. Chem. Phys.* **29**, 340 (1958).

53. J.E. Williams, Jr., V. Buss, L.C. Allen, P.v.R. Schleyer, W.A. Lathan, W.J. Hehre and J.A. Pople, *J. Am. Chem. Soc.* **92**, 2141 (1970).

54. T. Nishikawa, T. Itoh and K. Shimoda, *J. Chem. Phys.* **23**, 1735 (1955).

55. E.V. Ivash and D.M. Dennison, *J. Chem. Phys.* **21**, 1804 (1953).

56. J.E. Griffiths, *J. Chem. Phys.* **38**, 2879 (1963).

57. D.R. Lide, Jr., and D.E. Mann, *J. Chem. Phys.* **27**, 868 (1957).
58. R.W. Kilb, C.C. Lin and E.B. Wilson, *J. Chem. Phys.* **26**, 1695 (1957).
59. M. Rabinovitz and A. Pines, *J. Am. Chem. Soc.* **91**, 1585 (1969).
60. M.A. Robb and I.G. Csizmadia, *J. Chem. Phys.* **50**, 1819 (1969).
61. A. Veillard, *Theor. Chim. Acta* **5**, 413 (1966).
62. T. Kasuya and T. Kojima, *J. Phys. Soc. Japan* **18**, 364 (1963).
63. V. Kaldor and I. Shavitt, *J. Chem. Phys.* **44**, 1823 (1966).
64. W.E. Palke and R.M. Pitzer, *J. Chem. Phys.* **46**, 3948 (1967).
65. R.H. Hunt, R.A. Leacock, C.W. Peters and K.T. Hecht, *J. Chem. Phys.* **42**, 1931 (1965).
66. A. Rauk, S. Wolfe and I.G. Csizmadia, *Can. J. Chem.* **47**, 113 (1969).
67. K.S. Pitzer, *J. Chem. Phys.* **5**, 473 (1937).
68. A. Veillard, unpublished results.
69. J.L. Margrave, M.A. Frisch, R.G. Bautista, R.L. Clark and W.S. Johnson, *J. Am. Chem. Soc.* **85**, 546 (1963).
70. F.A.L. Anet and A.J.R. Bourn, *J. Am. Chem. Soc.* **89**, 760 (1967).
71. G.G. Engerholm, A.C. Luntz, W.D. Gwinn and D.D. Harris, *J. Chem. Phys.* **50**, 2446 (1969).
72. J. Higuchi, *J. Chem. Phys.* **24**, 535 (1956).
73. H. Kaplan, *J. Chem. Phys.* **26**, 1074 (1957).
74. R. Moccia, *J. Chem. Phys.* **40**, 2176 (1964).
75. B.D. Joshi, *J. Chem. Phys.* **43**, S 40 (1965).
76. V. Kaldor and I. Shavitt, *J. Chem. Phys.* **45**, 888 (1966).
77. D.M. Bishop, *J. Chem. Phys.* **45**, 1787 (1966).
78. B. Tinland, *Chem. Phys. Letters* **2**, 433 (1968).
79. J.D. Swalen and J.A. Ibers, *J. Chem. Phys.* **36**, 1914 (1962).
80. H. Basch, M.B. Robin and N.A. Kuebler, *J. Chem. Phys.* **47**, 1201 (1967).
81. M.A. Robb and I.G. Csizmadia, *Theor. Chim. Acta* **10**, 269 (1968).
82. C.C. Costain and J.M. Dowling, *J. Chem. Phys.* **32**, 158 (1960).
83. D.T. Clark, *Chem. Commun.* **1969**, 637.
84. A. Mannschreck, J. Linss and W. Seitz, *Ann.* **727**, 224 (1969).
85. F.A.L. Anet, J.C. Jochims and C.H. Bradley, *J. Am. Chem. Soc.* **92**, 2557 (1970).
86. D.M. Rutledge and A.F. Saturno, *J. Chem. Phys.* **43**, 597 (1965).
87. G.E. Hansen and D.M. Dennison, *J. Chem. Phys.* **20**, 213 (1952).
88. R.L. Redington, W.B. Olson and P.C. Cross, *J. Chem. Phys.* **36**, 1311 (1962).
89. W.S. Benedict and E.K. Plyler, *Can. J. Phys.* **35**, 1235 (1957).
90. C.A. Burrus, Jr., A. Jache and W. Gordy, *Phys. Rev.* **95**, 900 (1954).
91. R.M. Stevens, *J. Chem. Phys.* **52**, 1397 (1970).
92. A. Veillard and J. Demuynck, *Chem. Phys. Letters* **4**, 476 (1960).
93. M.E. Schwartz, *J. Chem. Phys.* **51**, 4182 (1970).
94. G. Winnewisser, M. Winnewisser and W. Gordy, *J. Chem. Phys.* **49**, 3748 (1968).
95. P.A. Kollman and L.C. Allen, *Chem. Phys. Letters* **5**, 75 (1970).
96. D. Coffey, C.O. Britt and J. Boggs, *J. Chem. Phys.* **49**, 591 (1968).
97. P. Millié, private communication.
98. R.W. Kilb, C.C. Lin and E.B. Wilson, *J. Chem. Phys.* **26**, 1695 (1970); D.R. Herschbach, *J. Chem. Phys.* **31**, 91 (1959).
99. A.C. Hopkinson, K. Yates and I.G. Csizmadia, *J. Chem. Phys.* **52**, 1784 (1970).
100. T. Miyazawa and K.S. Pitzer, *J. Chem. Phys.* **30**, 1076 (1959).
101. H. Hogeveen, A.F. Bickel, C.W. Hilbers, E.L. Mackor and C. MacLean, *Rec. Trav. Chim.* **86**, 687 (1967).

155

102. D.T. Clark and D.M.J. Lilley, *Chem. Commun.* **1970**, 603.
103. (a) S. Wolfe, A. Rauk, L.M. Tel and I.G. Csizmadia, *Chem. Commun.* **1970**, 96;
      (b) S. Wolfe, A. Rauk and I.G. Csizmadia, *J. Am. Chem. Soc.* **91**, 1567 (1969).
104. E. Zeeck, *Theor. Chim. Acta* **16**, 155 (1970).
105. J.B. Robert, H. Marsman and J.R. Van Wazer, *Chem. Commun.* **1970**, 356.
106. J.S. Wright and L. Salem, *Chem. Commun.* **1969**, 1370.
107. A. Rauk, L.C. Allen and K. Mislow, to be published.
108. A. Rauk, L.C. Allen and E. Clementi, *J. Chem. Phys.* **52**, 4133 (1970).
109. A. Pipano, R.R. Gilman, C.F. Bender and I. Shavitt, *Chem. Phys. Letters* **4**, 583 (1970).
110. C.F. Bender, *Theor. Chim. Acta* **16**, 401 (1970).
111. J.M. Lehn and B. Munsch, *Chem. Commun.* **1970**, in press.
112. J.M. Lehn and B. Munsch, *Chem. Commun.* **1970**, 994.
113. J.B. Moffat and C. Vogt, *J. Mol. Spectrosc.* **33**, 494 (1970).
114. W.H. Fletcher and F.B. Brown, *J. Chem. Phys.* **39**, 2478 (1963); T.R. Jones and N. Sheppard, *Chem. Commun.* **1970**, 715; J.K. Tyler, private communication.
115. K. Morokuma, I. Pedersen and M. Karplus, *J. Chem. Phys.* **48**, 4801 (1968).
116. B. Levy and M.C. Moireau, to be published.
117. C. Thomson and A. Veillard, private communication.
118. F.A. Claxton and N.A. Smith, *J. Chem. Phys.* **52**, 4317 (1970).
119. A. Rauk, L.C. Allen and K. Mislow, unpublished results.
120. T. Ueda and T. Shimanouchi, *J. Chem. Phys.* **49**, 470 (1968).
121. R.J. Buenker, *J. Chem. Phys.* **48**, 1368 (1968).
122. L.J. Schaad, L.A. Burnelle and K.P. Dressler, *Theor. Chim. Acta* **15**, 91 (1969).
123. J.M. André, M.-C. André and G. Leroy, *Chem. Phys. Letters* **3**, 695 (1969).
124. B. Roos and P. Siegbahn, *Theor. Chim. Acta* **17**, 199 (1970).
125. A. Pipano, R.R. Gilman and I. Shavitt, *Chem. Phys. Letters* **5**, 285 (1970).
126. A. Rauk, L.C. Allen and K. Mislow, *Angew. Chem.* **82**, 453 (1970).
127. J.M. Lehn, "Topics in Current Chemistry," Vol. 15, p. 311, Springer, Heidelberg, 1970.
128. J.B. Lambert, "Topics in Stereochemistry," Wiley (Interscience), New York, to be published.

Conformational Studies of Nitrogen Heterocycles

Robert E. Lyle, John J. Thomas and David A. Walsh
Department of Chemistry
University of New Hampshire
Durham, New Hampshire

It is a necessary but not a sufficient requirement that the hydrogens of a methylene group be diastereotopic in order for them to appear as nonequivalent in the nuclear magnetic resonance (nmr) spectrum (*1*). If the magnetic environments of the two hydrogens are not particularly different, as is the case if the chirality is far removed from the geminal hydrogens, the diastereotopic hydrogens will appear as a singlet resonance band (*2*). A better understanding of the structural requirements which lead to observable nonequivalence of geminal protons should permit this physical measurement to be used as a probe for conformational as well as configurational chirality.

The investigation of the nmr spectra of a series of *N*-benzylpiperazine derivatives led to the formulation of one conformational relationship with observable nonequivalence of the methylene protons of benzyl substituents. An *N*-benzyl group attached to a six-membered ring adjacent to a chiral center will exhibit an AB quartet for the methylene protons in the nmr spectrum if the substituent is equatorial, but not if the substituent is axial (*3*). This relationship was deduced from the spectra of the piperazines and piperidines listed in Fig. 1.

The steric relationship of the phenyl group of the benzyl substituent and the equatorial ring substituent suggests that there would be an unequal population of the three rotomers of the benzyl group. In the energetically favored conformer, the methylene protons of the benzyl group will be situated as if they were axial and equatorial hydrogens at the 2 position of a piperidine (Fig. 2). The resonance signals for these protons have been shown to differ in chemical shift by more than 1 ppm (*4*). The cause of this chemical-shift difference is the object of controversy, but its existence is accepted (*5, 6*).

157

Starting with the hypothesis developed previously (3), a number of nmr results may be related to conformational arguments. *cis*-1-Benzyl-2,5-dimethylpiperazine (I) would be expected to exist in a mobile conformational equilibrium (Fig. 3). Any free-energy difference between the two conformers

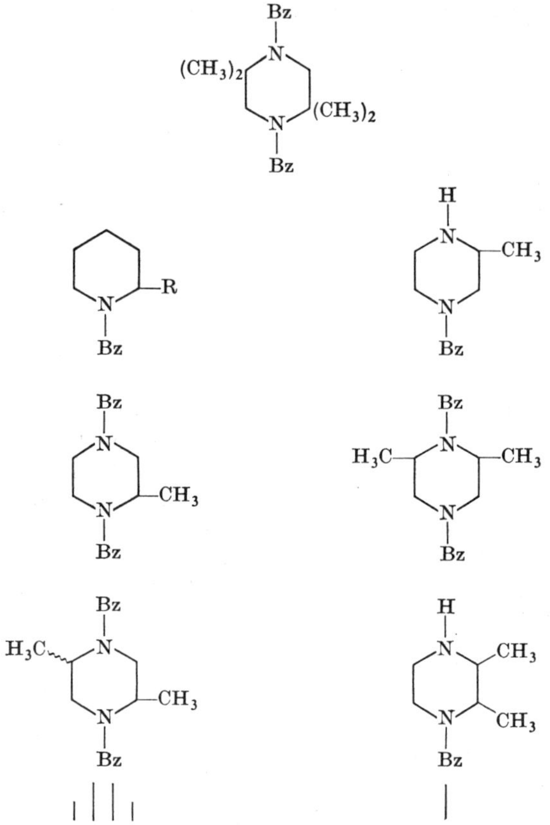

Figure 1   The methylene protons of the benzyl groups of the compounds at the left appear as AB quartets in the nmr spectra, while those at the right and at the center give resonance signals which appear as singlets.

would result only from an interaction of the benzyl group with the adjacent methyl substituent. Since the resonance band for the benzyl protons is a singlet, this suggests that the steric interaction and entropy difference are sufficient to require that the conformational equilibrium strongly favor the 2 axial methyl in which the two benzylic rotomers are approximately of equal importance in the rotomeric equilibrium. The accidental magnetic equivalence of the benzylic protons was maintained in nmr spectra determined neat, in

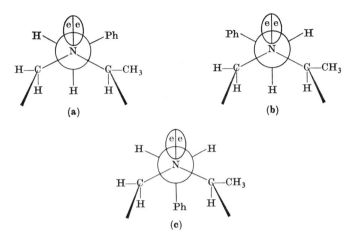

Figure 2   The three rotomers of the benzyl–N bond which should represent energy
minima. Rotomer (c) should contribute little and rotomer (b) should be of
considerably lower energy than rotomer (a).

deuteriochloroform, benzene and pyridine. Except for broadening of the
spectral bands, little change in the spectrum was noted on cooling the deuteriochloroform solution to −60°C. In methanol solution there was a small but
definite break of the benzyl peak into an AB quartet. On cooling this solution,
the position of the resonance band shifted upfield but showed no further
separation into a quartet. These changes are consistent with a decrease in the
free-energy difference between the ring conformers resulting from solvation
by methanol of the secondary function.

Conversion of the secondary amine of I to the amide function should completely change the energetics of the ring conformational equilibrium, for
serious steric interaction between the amide group and an equatorial substituent has been demonstrated by Johnson (7) and Paulsen (8). Such an
interaction should cause the ring conformer of II, in which the methyl group
adjacent to the amide function is axial, to be heavily favored energetically.

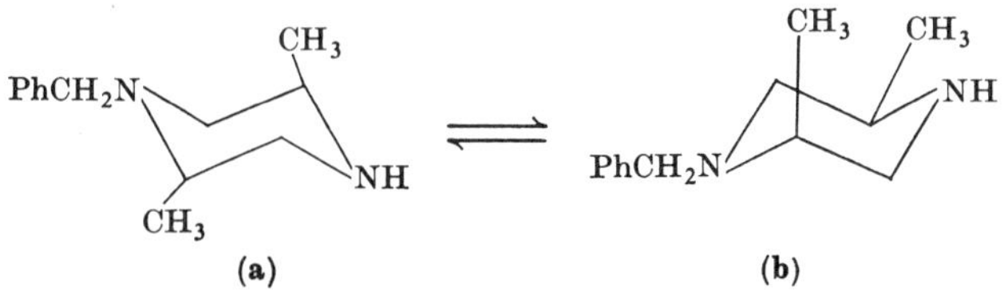

Figure 3   The ring conformers of *cis*-1-benzyl-2,5-dimethylpiperazine (I). Conformer (b)
is probably of lower energy than (a) due to the eq-methyl:benzyl interaction
in (a).

This would require the methyl adjacent to the benzyl group in this *cis* isomer to be equatorial. The nmr spectra of 4-acetyl-, 4-benzoyl- and 4-(4-dibenzfuroyl)-1-benzyl-*cis*-2,5-dimethylpiperazines (II) all showed resonance signals for the benzylic methylene protons as AB quartets with a chemical-shift separation of more than 1 ppm (Fig. 4).

cis-1,4-Dibenzyl-2,5-dimethylpiperazine (III) should exist as a conformational equilibrium of two identical ring conformers. The methylene groups exchange roles in the two conformers, however, for the methylene which is adjacent to an equatorial methyl group in one conformer is adjacent to an

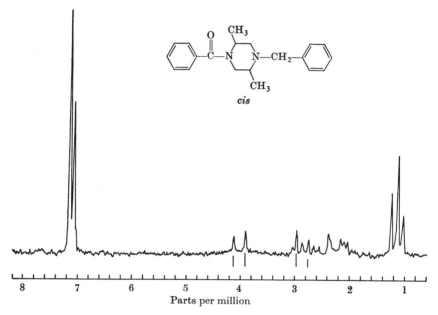

Figure 4 Nuclear magnetic resonance spectrum of 4-benzoyl-1-benzyl-*cis*-2,5-dimethylpiperazine (II) illustrating the AB quartet shown by all amides of *cis*-1-benzyl-2,5-dimethylpiperazine (I).

axial methyl group in the other. If the interconversion of the two conformers is rapid, as would be expected at room temperature, the benzylic methylene protons should be in approximately identical magnetic environments 50% of the time and in nonequivalent environments 50% of the time. As would be predicted, this condition led to an nmr spectrum in which the benzylic protons appeared as an AB quartet with a chemical-shift difference 50% of that observed with the *trans* isomer (IV).

As the solution of *cis*-1,4-dibenzyl-2,5-dimethylpiperazine (III) is cooled, the interconversion of the ring conformations should be slowed and two kinds of benzylic protons should become evident. At −81°C the original AB quartet of $\Delta\delta = 0.6$ ppm collapsed into an AB quartet of chemical-shift difference of 1.2 ppm, and a singlet appeared which was located in the center of the quartet (Fig. 5). These observations are exactly consistent with the slowing of ring inversion.

The results of the spectral studies with *cis*-1,4-dibenzyl-2,5-dimethyl-piperazine (III) suggested that the chemical-shift separation of the AB

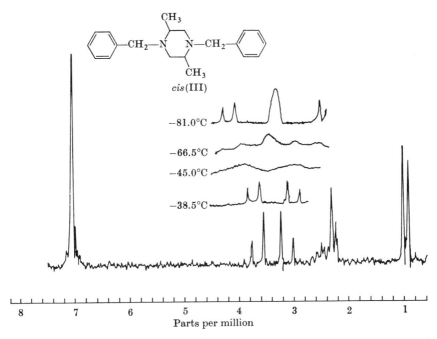

Figure 5   The change in the nmr signal for the methylene protons of the benzyl groups of *cis*-1,4-dibenzyl-2,5-dimethylpiperazine on cooling the sample.

quartet might provide a quantitative estimate of the amount of the ring conformer with a vicinal equatorial substituent in the equilibrium of ring conformers. In the unlikely event that the nature of the aromatic ring of the benzylic group was unimportant to the chemical-shift difference of the AB quartet, this approach would provide a convenient and sensitive method for comparing the effective sizes of substituents. Thus the chemical-shift differences of the resonance signals of the methylene protons of compounds such as V could provide an estimate of the free-energy differences of the ring conformers and a comparison of the magnitude of steric interactions of Ar and phenyl with a vicinal methyl group.

cis - 1 - Benzyl - 4 - (1 - dibenzofuranylmethyl) - 2,5-dimethylpiperazine (Va) showed two superimposed AB quartets. The quartet in the spectrum of Va with lines at the same chemical shifts as those for *cis*-1,4-dibenzyl-2,5-dimethylpiperazine (III) gave proton signals separated by 34.0 Hz, while the

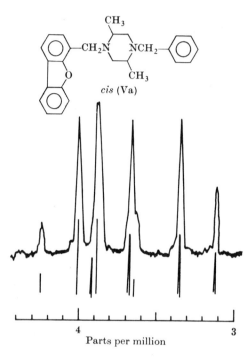

Figure 6   Nuclear magnetic resonance signals for the methylene protons of the *cis*-1-benzyl-4-(1-dibenzofuranylmethyl)-2,5-dimethylpiperazine (Va).

dibenzyl derivative III showed a separation of 35.5 Hz. If it is assumed that the latter separation corresponds to 50% equatorial methyl group adjacent to the benzylic substituent, then in V the conformer ratio would be 47.9:52.1, corresponding to a free-energy difference of the conformers of about 50 cal favoring i. If this approach is valid, it provides a method for estimating very small free-energy differences not detectable by most methods.

The unequal distribution of rotomers of a benzylic substituent adjacent to an equatorial substituent provides the basis for a convenient method for distinguishing between configurational isomers if the conformational analysis is obvious. For example, the reaction of benzylamine with 1,1,2,5-tetramethyl-4-oxopiperidinium iodide (VI) produces a mixture of 1-benzyl-2,5-dimethyl-4-piperidones (VII) (9). The examination of the nmr spectrum of the purified

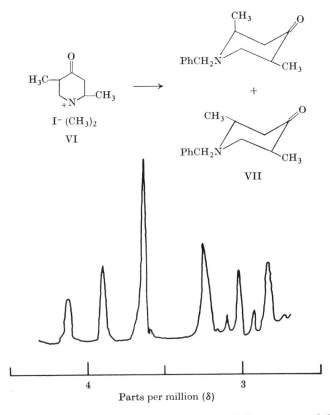

Figure 7   Nuclear magnetic resonance signals for the methylene protons of the benzyl group in the product mixture containing *cis*- and *trans*-1-benzyl-2,5-dimethyl-4-piperidone (VII).

major fraction shows an AB quartet for the benzylic protons (*10*). This is a clear indication that the major product is *trans*-1-benzyl-2,5-dimethyl-4-piperidone (VII).

An understanding of the structural features which remove accidental magnetic equivalence of diastereotopic hydrogens of *N*-benzyl substituents in six-membered heterocycles provides a convenient method for studying the ring conformation. Preliminary investigation has shown this procedure to be useful for determining configurational differences of disubstituted rings and for estimating vicinal, steric interactions.

ACKNOWLEDGMENTS

The authors wish to express appreciation to Drs. J.J. Uebel and Morton Raban for discussions concerning this research. The financial assistance of the National Cancer Institute of the National Institutes of Health and the grantors of the Petroleum Research Fund of the American Chemical Society is gratefully acknowledged.

REFERENCES

1. K. Mislow and M. Raban, *in* "Topics in Stereochemistry," Vol. 1, pp. 22–37, N. Allinger and E. Eliel, eds., Wiley (Interscience), New York, 1967.
2. R.K. Hill and T.H. Chan, *Tetrahedron* **21**, 2015 (1965).
3. R.E. Lyle and J.J. Thomas, *Tetrahedron Letters* **1969**, 897.
4. H.P. Hamlow, S. Okuda and N. Nakagawa, *Tetrahedron Letters* **1964**, 2553.
5. J.B. Lambert, R.G. Keske, R.E. Carhart and A.P. Jovanovich, *J. Am. Chem. Soc.* **89**, 3761 (1967); J.B. Lambert and R.G. Keske, *Tetrahedron Letters* **1969**, 2023.
6. M.J.T. Robinson, *Tetrahedron Letters* **1968**, 1153.
7. R.A. Johnson, *J. Org. Chem.* **33**, 3627 (1968).
8. H. Paulsen and K. Todt, *Chem. Ber.* **100**, 3385 (1967).
9. E.A. Mistryukov, N.I. Aronova and V.F. Kucherov, *Izv. Akad. Nauk SSSR, Ser. Khim.* **1961**, 932.
10. A.F. Casy, University of Alberta, Canada, private communication, 1969.

Energetics of Isomeric Transition States and Competitive Reaction Pathways in Conformational Analysis

James McKenna
Chemistry Department
The University
Sheffield, England

The reactivity of conformationally mobile systems is difficult to treat quantitatively, but schematic qualitative discussions are readily based on transition-state theory, modified as required to take into account the dynamic character of chemical reactions. The energy schematics of two different types of conformationally mobile systems (one of which in particular has caused difficulties for many authors) and some related qualitative interpretations form the subject of this chapter.

For a conformationally mobile system $1 \rightleftharpoons 2 \rightleftharpoons \cdots$, with fast interconversions between conformers and with each conformer present in at least moderate proportions, a physical property, p, or rate constant, k, is often written as

$$p = N_1 p_1 + N_2 p_2 + \cdots \tag{1}$$

$$k = N_1 k_1 + N_2 k_2 + \cdots \tag{2}$$

where N_1, N_2, ..., are the fractions of each conformer at equilibrium, and p_1, p_2, ..., k_1, k_2, ..., are values for analogous conformationally "rigid" or strongly biased systems.

These empirical equations have been the basis of several well-known procedures for quantitative conformational analysis (1), although they cannot be justified theoretically because of the differential operation of a host of polar, steric, magnetic, and other effects. A more specific criticism is sometimes (2) leveled at the kinetic procedure as used for the determination of conformational equilibria in cyclohexanes, e.g., by use of the reaction cyclohexyl-R $\rightarrow$

cyclohexyl-R′ via the transition states cyclohexyl-R⁺ in Fig. 1. The utilization of *two* transition states, A⁺ and B⁺ (cyclohexyl-R⁺ in both possible chair conformations) is called into question; these transition states would correspond to those of the reactions of the *cis*- and *trans*-4-*t*-butyl models.

In fact, however, a more pertinent analytical point of difference is that the strongly biased diastereomers use essentially only one transition state each (neglecting boat forms), but *each* conformer of the parent mobile system uses *both* transition states A⁺ and B⁺, in ratio $1:\exp-\Delta G_{\mathrm{ts}}^{*}/RT$, where $\Delta G_{\mathrm{ts}}^{*}$ is the free-energy difference between the competitive transition states; this is a

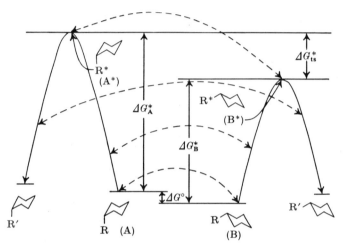

Figure 1   Transition-state diagram for reaction of a conformationally mobile cyclohexyl system.

fairly small ratio if $\Delta G_{\mathrm{ts}}^{*}$ is small, as is ΔG° in such applications. This striking difference does not in itself affect the applicability of equation (2), since removal of all the interconversion pathways between species or competitive energetic surfaces (represented by curved dashed lines with double arrows in Fig. 1) except that between the reactant conformers A and B, does not affect the rate of passage through either transition state: provided A and B are at equilibrium, so also, in transition-state theory, are A⁺ and B⁺. The analytical situation therefore is that the system reacts *as if* all A were transformed into product via A⁺, and all B via B⁺. Satisfactory application of equation (2) requires merely that $\Delta G_{\mathrm{A}}^{*}$ and $\Delta G_{\mathrm{B}}^{*}$ should be close to the corresponding free energies of activation for the models chosen, usually the *cis*- and *trans*-4-*t*-butyl analogs.

The question of the relative extent of utilization of two transition states in the case represented in Fig. 1 clearly involves consideration of the magnitude of ΔG_{ts}^{*}, the equatorial preference,* so to speak, of the "group" R^{+}. This sort of problem has been generally neglected, although in other areas of conformational analysis, e.g., unimolecular conformational inversions, transition-state energetics have been extensively scrutinized (3). Such considerations are also required for rationalization of the differential reactivities at axial and epimeric equatorial groups in strongly biased or conformationally "rigid" systems such as steroids: the relative signs and magnitudes of ΔG° (for the appropriate *configurational* equilibrium in such cases) and ΔG_{ts}^{*} will determine whether the

Table I   Reactions of Derivatives of 5α-Cholestane

Reaction[a]	Conditions	Relative rate constants	
3α-NMe$_2$ + MeI	Nitrobenzene–benzene at 0°C	1	150:1
3β-NMe$_2$ + MeI	Nitrobenzene–benzene at 0°C	150	
3α-CH$_2$NMe$_2$ + MeI	Nitrobenzene–benzene at 0°C	19	0.8:1
3β-CH$_2$NMe$_2$ + MeI	Nitrobenzene–benzene at 0°C	15	
3α-CO$_2$Me* + OMe$^-$	Dioxan–methanol at 3.5°C	1.0	3.7:1
3β-CO$_2$Me* + OMe$^-$	Dioxan–methanol at 3.5°C	3.7	
3α-CH$_2$CO$_2$Me* + OMe$^-$	Dioxan–methanol at 3.5°C	2.9	1.3:1
3β-CH$_2$CO$_2$Me* + OMe$^-$	Dioxan–methanol at 3.5°C	3.9	
3α-OAc* + OH$^-$	Dioxan–water–methanol at 6°C	1.0	2.2:1
3β-OAc* + OH$^-$	Dioxan–water–methanol at 6°C	2.2	

[a] The asterisks denote ^{14}C labels.

reaction for an equatorial group will be faster or slower than for an epimeric axial group, and by how much. We now discuss examples from our own recent work in this area.

We sought an adequate interpretation of one of the classical observations in conformational analysis, that acetyl derivatives of axial alcohols are usually saponified more slowly than are the derivatives of the epimeric equatorial alcohols. It is by no means a priori obvious why this should be so: vague suggestions that differential ease of attack of base on carbonyl carbon (two bond lengths from the ring) is the deciding feature seemed quite unsatisfactory. The relative rate constants (4) for the Menschutkin reactions support this criticism, as do the values (Table I) (5) for exchange of ^{14}C-labeled methoxyl

* We assume in Fig. 1 that the transition states have chair conformations, but a similar analysis can of course be made if they have not.

in the methyl esters from 3α- and 3β-carboxymethyl-5α-cholestane where base attack (as in epimeric O-acetyl derivatives) is in fact taking place at carbonyl–carbon two bond lengths from the ring, but where the rate-constant ratio is nevertheless much nearer unity than is the ratio (5) for hydrolysis of 3α- and 3β-acetoxy-5α-cholestanes run under conditions as near as experimentally convenient to those used in the methoxyl-exchange reactions. (The difference in rate constants for methoxyl exchange with the methyl esters of the epimeric 3-carboxylic acids (5) is larger still.*) All these reactions would be expected to proceed via the usual addition intermediate, tetrahedral at the former acyl carbon atom, whose formation is probably rate-determining for the reaction

Table II   Reactions of Basic Derivatives of 5α-Cholestane
with 2,4-Dinitrochlorobenzene[a]

R–NH$_2$ + 2,4-DNCB	Rate-constant order $3\alpha \gtrapprox 6\beta > 3\beta > 6\alpha$ Overall range ~2:1
R–NHMe + 2,4-DNCB	Rate-constant order $6\alpha > 3\beta > 3\alpha > 6\beta$ Overall range ~20:1
Formation of R–$\overset{+}{N}H_2$–[PhCl(NO$_2$)$_2$]$^-$ apparently rate-determining: no acetate catalysis	

[a] In 98.5% ethanol at 99.5°C.

under observation. Such intermediates have formal negative charges on the acyl–oxygen atoms, but, by induction, these will be partly shared by the alkyl–oxygens. It seems that the alkyl–oxygen atoms must somehow be involved in the interpretation of the relative hydrolysis rates of epimeric steroidal O-acetyl derivatives, and we suggest that the different steric strains hindering solvation at these atoms in the epimeric transition states provides the best qualitative interpretation.

In the next example (Table II), we consider the relative rates of reaction of a series of primary cholestanylamines with 2,4-dinitrochlorobenzene under conditions (absence of acetate catalysis) indicating that formation of the N-tetrahedral intermediate R–$\overset{+}{N}H_2$–[PhCl(NO$_2$)$_2$]$^-$ is rate-determining (6). The order of rate constants is somewhat unexpected, since bulk is being added

* We are not aware of a similar overall comparison in the literature, but individual ratios quoted for comparable epimeric pairs of O-acetyl derivatives or methyl carboxylates are usually higher than the above values, indicating that the experimental conditions we used are such that all ratios are kept low.

at the nitrogen atoms directly attached axially or equatorially to the ring system. In the quaternization of cholestanyldimethylamines with methyl iodide, equatorial bases react faster than their axial epimers except in special structural circumstances (4b, 7) and the reactivity order (6) for cholestanyl-methylamines with dinitrochlorobenzene (Table II) is also fairly "normal." If we follow the line of analysis suggested above, we may say that the equatorial preference for the nitrogenous group in the transition state leading to $-\overset{+}{N}H_2-[PhCl(NO_2)_2]^-$ is *smaller* than that for $-NH_2$. One cannot presently give a reasonably assured qualitative interpretation of this apparent fact, but the parallel with the order (8) of equatorial preferences $-NH_2 > NHMe$ in aprotic

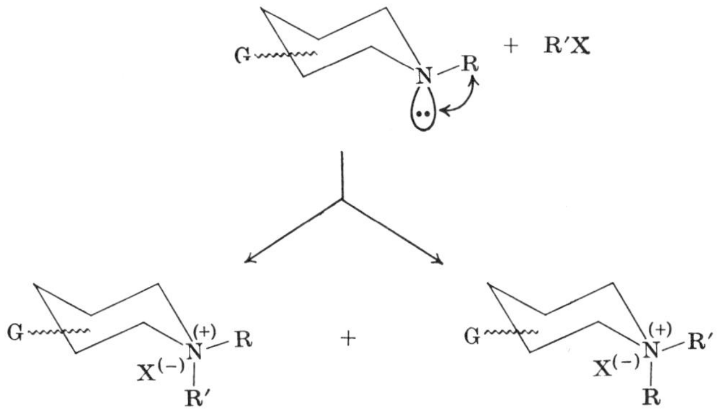

Figure 2 Diastereomeric quaternary salts from N-alkylpiperidines (G = biasing or reference group or fused ring).

solvents is intriguing, and one would very much like to know (a) what would be the relative equatorial preferences for these two groups and similar pairs in protic solvents, and more particularly, the order for the N-protonated salts,* (b) at what fairly precise stage in the dinitrophenylation mechanism is the proton actually lost from the putative N-tetrahedral intermediates, and (c) what other reactions of epimeric pairs of alicyclic primary amines would also show the "wrong" reactivity order (9).

We now consider a system where a conformationally mobile reactant yields two *diastereomeric* products: a suitable substituted tertiary piperidine, in principle and usually also in practice, gives a mixture of two quaternary salts (Fig. 2) when a different N-alkyl group (R') to that initially present (R) is introduced in the quaternization. The energetics and reaction pathways in this system are shown in Fig. 3. No reversible conversion between competitive

---

* In benzene, the same relative rate-constant order for the reaction of 5α-cholestan-3α- and 3β-ylamines with 2,4-dinitrochlorobenzene is observed (6) as in alcohol; the 6-epimers have not yet been examined in this solvent.

products or reaction surfaces is usually possible here (see Fig. 1) during the reaction. However, either reactant conformer may be transformed into the product stereochemically analogous to the other (schematically represented by the straight dashed lines with arrows pointing one way)* or, naturally,

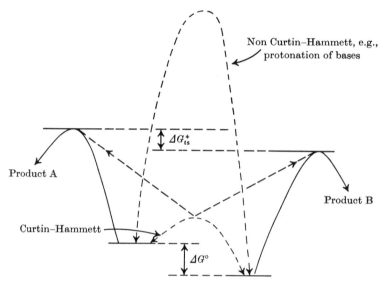

Figure 3 Kinetic analysis of competitive quaternizations.

$$\Delta G_{ts}^{\ddagger} \approx \Delta V_{ts}^{\ddagger} = 2.3\,RT\log_{10} P, \tag{a}$$

where P is the product ratio.

$$k_{obs} = k_e\,N_e + k_a\,N_a \tag{b}$$

and

$$P = k_e\,N_e/k_a\,N_a \quad\text{or}\quad k_a\,N_a/k_e\,N_e,$$

where subscripts e and a refer to equatorial and axial conformers, respectively, k's are rate constants, and N's are fractions.

into the product stereochemically analogous to itself (curved solid lines with arrows pointing one way). The reactant conformers with equatorial and axial N-alkyl groups are, of course, themselves readily interconverted. The analytical result (as before with Fig. 1) is that the system reacts *as if* each base conformer were converted only into the stereochemically analogous product; hence, equations (b) of Fig. 3 are, apparently, valid. However, it is incorrect to make such (commonly made) statements as, "the quaternization product is derived

* What this means is that it is possible for the base conformer with the group R, say, equatorial, to be attacked by the quaternizing agent R'X in such a way that the quaternary salt with R axial and R' equatorial is formed without initial conformational inversion at nitrogen in the reactant. The inversion may take place *during* the reaction, so there are two separate reaction pathways probably meeting, as indicated in the figure, before the transition state is reached.

mainly from the more stable base conformer." *Each* product is in fact derived from *both* conformers in equal proportions.

Since the energetic barrier separating the conformers is usually expected to be much lower than those separating conformers from products, the Curtin-Hammett principle holds (equation (a), Fig. 3). The "non-Curtin-Hammett" energetics (Fig. 3) would apply, for example, to fast protonation of the base, but in any attempts to make use of this fact (*10*) for measurement of the reactant–conformer ratio (from the ratio of diastereomeric proton salts observed in a strongly acidified solution) allowance will have to be made for the degree of partial equilibration of the salts by reversible proton-transfer *during* the mixing process, and this is by no means easy to do when two liquids are mixed.

Our own interest (*11*) in this system has been in the stereospecificities and stereoselectivities of the quaternizations: which of the two possible isomeric products predominate in different quaternizations, and by how much? These are, of course, essentially structural and analytical queries, but so far as the competitive kinetics are concerned any interpretation of an observed product ratio P on transition-state theory must be in terms of ΔG_{ts}^{*} (equation (a), Fig. 3). Conversely, substantial variations in product ratios of related reactions may be expected to provide a clue to the stereospecificities of the reactions, following qualitative consideration of the energetics of the appropriate competitive pairs of transition states.* However, a noteworthy feature of the literature of the field has been the reluctance of some other groups to use the Curtin-Hammett principle in this most obvious application; in the main this appears to have stemmed from a simultaneous preoccupation with the related (and important) problem of conformational equilibria in the reactant bases. Thus Fodor and his collaborators initially (*13*) took the extreme and erroneous view that the quaternization product ratios directly reflected (in effect) ΔG° for the appropriate base-conformer conversion rather than ΔG_{ts}^{*}, and many other workers followed the same line of thought. More recently (*14*), Katritzky and his collaborators set themselves the extremely difficult† task of deciding the stereospecificity of quaternizations of N-alkylpiperidines by using equations (b) of Fig. 3. It seems doubtful if this approach could have been effective, but some critical experimental errors (*11*) in the work and in earlier related literature prevents a full assessment of the point.

In our own work (*11*) we took the very simple and, it turned out, effective

* We are not, however, happy about the interpretation of quite small variations in product ratios on the basis of the Hammond postulate (*12*).

† The appropriate base-conformer equilibrium constants in tertiary piperidines are very difficult to determine satisfactorily, and the problems of finding suitable model bases in rigid or biased systems with rate constants comparable to those of the conformers in conformationally mobile piperidine systems are of necessity even more acute than are the analogous problems of cyclohexane chemistry.

view that if there were marked differences in stereoselectivity between the reactions $>N$Alk + MeI (Alk being a primary alkyl group other than, and therefore larger than, methyl) and $>N$Me + AlkI, when the reactions were run under comparable conditions, the former reaction should favor increased axial alkylation, and the latter increased equatorial alkylation. Hence we deduced a preference for axial methylation in the first four examples shown in Table III, while alkylation of N-methyl bases gives near-zero stereoselectivity in the first three cases, but a preference for axial attack in the fourth (the camphidine system). In the next example (the 4-aza-5α-cholestane system), we have near-zero stereoselectivity for methylation, and a preference for equatorial alkylation of the N-methyl base. In the tropane system, for most pairs of reactions, we observe very similar stereoselectivity patterns for introduction of methyl or primary alkyl groups in one or the reverse order, so no deductions regarding stereospecificity for such reactions can be made. However, the ratios for formation of mixtures of N-methyl-N-isopropyl salts so strongly suggested preferred axial quaternization that mainly on this evidence alone we were led (*11*) to question for the first time the correctness of the classical conclusions of Fodor and his collaborators (*15*) that tropines (3-hydroxy-tropanes) undergo preferred equatorial quaternization, although these conclusions were apparently backed by both chemical and x-ray evidence. Repetition of the earlier work by Bottini (*16*) and by Fodor (*17*) appeared, however, to suggest a preference for axial quaternization, although some experimental ambiguities in the recent work still remain (*17a*).

The 3-azabicyclo[3.2.1]octane system shows a preference for axial quaternization in three of the reactions indicated in Table III; unusually, reaction of the N-methyl base with ethyl iodide proceeds preferentially via equatorial attack (*18*). Finally, the figures for the 3-azabicyclo[3.3.1]-nonane system would also initially suggest a preference for axial quaternization, but, since the nuclear magnetic resonance (nmr) evidence indicates that quaternary salts in this system have a conformation with a piperidine boat, we reasoned (*19*) that this would also be true for the competitive transition states and so deduced preferred *boat–axial* quaternization, i.e., equatorial, from a configurational standpoint. This configurational conclusion was supported by subsequent independent structural work (*20*) by House and his collaborators.

The conclusions reached for all the above base systems are also supported by other work by our own group, based on analysis of nmr spectra (*21*), on earlier more empirical methods, and in two cases on x-ray analysis (*22*). The general observation is of preferred axial quaternization by methyl iodide in acetone, with a reduced tendency for this result in quaternizations with other alkyl iodides, so that in the latter case the stereoselectivity of the reaction may be near zero, or there may indeed be some preference for equatorial quaternization.

Table III Configuration by Product-Ratio Methods: Stereoselectivity of Quaternizations in Acetone[a]

Base system	$\rangle N$Alk + MeI[b]	$\rangle N$Me + AlkI[b]
(Ph-substituted ring, NR)	2–5:1	~1:1
(Me-substituted ring, NR)	2–3:1	~1:1
(ring, NR)	2–3:1	~1:1
(Me, Me—, Me-substituted ring, NR)	>12:1	1:2–3
(steroid, Me, Me, C$_8$H$_{17}$, N R)	1:1	2–5:1
(bicyclic, R/N)	2–5:1 >12:1 ($\rangle N$Pr^i)	1:2–5 1:2 (Pr^{i}I)
(bicyclic, NR)	>12:1 ($\rangle N$Et) ~6:1 ($\rangle N$CH$_2$Ph)	~2:1 (EtI) ~1:1.5 (PhCH$_2$I)
(bicyclic, NR)	8–10:1	1:2–3

[a] Argument: increased axial *or boat-axial* alkylation for $\rangle N$Alk + MeI; increased equatorial *or boat-equatorial* alkylation for $\rangle N$Me + AlkI.
[b] Alk = Et, Pr^n, Ph·CH$_2$–.

Faster than

Usually faster than

Figure 4 Degradation of diastereomeric quaternary salts (G = biasing or reference group or fused ring).

The relative ease of *removal* of axial or equatorial N-alkyl groups from quaternary salts has been found to parallel the stereospecificity of quaternization, i.e., the relative ease of *addition* of such groups. By suitable isotopic labeling experiments and otherwise (*23*), we have shown that axial N-methyl or N-benzyl groups are more readily removed by nucleophilic displacement with thiophenoxide anion than are the corresponding equatorial groups, and the same is also true for removal of N-ethyl groups by Hofmann elimination (Fig. 4). From the standpoint of consistency, it is certainly encouraging to find that each transition state on the left-hand side of Fig. 5 is of lower free energy than the corresponding transition state on the right, but quantitative evaluation of the respective free energies is, unfortunately, beyond the present scope of conformational analysis.

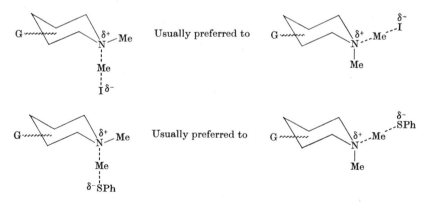

Figure 5 Correspondence in stereospecificity between quaternizations and quaternary-salt degradations (G = biasing or reference group or fused ring).

Qualitatively, however, we may note that in respect of space demands near the piperidine ring a partially attached (or detached) group (methyl, in our example) in the competitive transition states (Fig. 5) differs in two obvious ways from *the same* fully attached group: (a) it is less space-demanding because of the partial bond, longer and more flexible than a full bond, attaching it to the ring system, and (b) contrariwise, it is more space-demanding because of the arrangement of the C–H valences at methyl in the bipyramidal transition state. For methyl groups, the observed result is that the partly attached group is on balance in most base systems less space-demanding than the same fully attached group; this result could not have been predicted. By contrast, we should stress that the direction of change in the free energies of transition states associated with variation in the size and relative position of *different* exocyclic *N*-alkyl groups is quite simple to predict; as discussed above, such predictions provided a very successful method for deciding the preferred direction of quaternization of tertiary piperidines.

These examples emphasize the importance of careful qualitative consideration of the energetics of competitive reaction pathways and of isomeric transition states in discussions of reactivity in conformational analysis.

ACKNOWLEDGMENTS

The author acknowledges with gratitude the help of his research associates, and support from the Science Research Council, the Universities of Sheffield and Notre Dame, and Smith, Kline, and French Laboratories, Philadelphia.

REFERENCES

1. E.L. Eliel, N.L. Allinger, S.J. Angyal and G.A. Morrison, "Conformational Analysis," Wiley (Interscience), New York, 1965.
2. J.L. Mateos, C. Perez and H. Kwart, *Chem. Commun.* **1967**, 125.
3. G. Binsch, *in* "Topics in Stereochemistry," Vol. 3, pp. 97–192, E.L. Eliel and N.L. Allinger, eds., Wiley (Interscience), New York, 1968.
4. (a) P.B. Smith and J. McKenna, *Tetrahedron* **20**, 1933 (1964); (b) B.B. Gent and J. McKenna, *J. Chem. Soc.* **1956**, 573.
5. A.M. Farid, B.G. Hutley, J. McKenna and J.M. McKenna, unpublished work.
6. R.W. Horobin, B.G. Hutley, N.R. Khan and J. McKenna, *Chem. Commun.* **1966**, 636.
7. J. McKenna, J.M. McKenna, R. Ledger and P.B. Smith, *Tetrahedron* **20**, 2423 (1964).
8. J.A. Hirsch, *in* "Topics in Stereochemistry," Vol. 1, pp. 215–216, E.L. Eliel and N.L. Allinger, eds., Wiley (Interscience), New York, 1967; P.J. Brignell, K. Brown and A.R. Katritzky, *J. Chem. Soc.* (*B*) **1968**, 1462.
9. A.K. Macbeth, J.A. Mills and W.G.P. Robertson, *J. Chem. Soc.* **1951**, 2968; E.L. Eliel, E.W. Della and T.H. Williams, *Tetrahedron Letters* **1963**, 861.
10. H. Booth, *Chem. Commun.* **1968**, 802; J. McKenna and J. M. McKenna, *J. Chem. Soc.* (*B*) **1969**, 644; M.J.T. Robinson, Plenary Lecture, Brussels Symposium on Conformational Analysis, 1969.

11. D.R. Brown, R. Lygo, J. McKenna and J.M. McKenna, *J. Chem. Soc.* (*B*) **1967**, 1184 and earlier papers.
12. D.R. Brown, J. McKenna and J.M. McKenna, *Chem. Commun.* **1969**, 186.
13. G. Fodor, *Tetrahedron* **1**, 86 (1957); *Experientia* **11**, 129 (1955).
14. J. L. Imbach, A.R. Katritzky and R.A. Kolinski, *J. Chem. Soc.* (*B*) **1966**, 556.
15. G. Fodor, K. Koczka and J. Lestyán, *J. Chem. Soc.* **1956**, 1411; C.H. McGillavry and G. Fodor, *J. Chem. Soc.* **1964**, 597.
16. C.C. Thut and A.T. Bottini, *J. Am. Chem. Soc.* **90**, 4752 (1968).
17. G. Fodor, J.D. Medina and N. Mandava, *Chem. Commun.* **1968**, 581; G. Fodor, Lecture, Brussels Symposium on Conformational Analysis, 1969.
17a. G. Fodor, personal communication.
18. B.G. Hutley, J. McKenna and J.M. Stuart, unpublished work.
19. R. Lygo, J. McKenna and I.O. Sutherland, *Chem. Commun.* **1965**, 356.
20. H.O. House and C.G. Pitt, *J. Org. Chem.* **1966**, 1062.
21. D.R. Brown, J. McKenna and J.M. McKenna, *J. Chem. Soc.* (*B*) **1967**, 1195.
22. P.L. Jackson, M.Sc. Thesis, Sheffield, 1967; R. Brettle, D.R. Brown, J. McKenna and R. Mason, *Chem. Commun.* **1969**, 339.
23. J. McKenna, B.G. Hutley and J. White, *J. Chem. Soc.* **1965**, 1729; B.G. Hutley, J. McKenna and J.M. Stuart, *J. Chem. Soc.* **1967**, 1199; T. James, J. McKenna and J.M. Stuart, unpublished work.

Conformational Problems in Organic Hypervalent Molecules

Jeremy I. Musher
Belfer Graduate School of Science
Yeshiva University
New York, New York

The class of molecules which I refer to as hypervalent (*1*), i.e., those containing heteroatoms of Groups V–VIII of the periodic table in their higher valences, should provide much new chemistry in the years to come, if for no other reason than their general neglect over the years in which modern organic chemistry has developed. Among the most interesting features of hypervalent molecules are (a) the existence of two types of bonds from the heteroatom to the monofunctional ligand; (b) the unusual stereochemistry they introduce into organic structures by using their 90, 180 and 120° bond angles; and (c) their ability to isomerize or undergo intramolecular rearrangements, which makes their study a proper subject for the present symposium even if peripheral to the principal topic of hydrocarbon conformational analysis. Although I have discussed the subject of hypervalent molecules at some length recently (*1*), I will review the main ideas briefly here—restricting myself to the case of monofunctional ligands—before going on to a discussion of some conformational problems and an indication of some of our very recent experimental work. The type of problem I am concerned with has been discussed in reviews by Wittig (*2*) and Muetterties (*3*)—and indeed whatever experimental interest we have in the subject today is due to the impetus of their work—and by Schmutzler (*4*), Hellwinkel (*5*), Ramirez (*6*) and Westheimer (*7*). I should also note the recent extensive algebraic and topological analyses by Muetterties (*8*) and by Lauterbur and Ramirez (*9*) of the configurational rearrangements involved and I will later refer to some of the assumptions on which these analyses are implicitly based.

I. HYPERVALENT MOLECULES

I consider the bonding in molecules using heteroatoms of Groups V–VIII to be best described in terms of bond orbitals which utilize only s and p orbitals in the ground configuration and involve no d orbitals whatsoever. The types of bonds which involve p orbitals are the following:

(1) Ordinary or covalent single bonds in which a singly occupied p orbital from the heteroatom is bonded to a singly occupied atomic orbital from the ligand with

$$\psi = \psi_H + \lambda\psi_L$$

(2) Hypervalent bonds in which a doubly occupied p-orbital from the heteroatom is bonded with *two* singly occupied atomic orbitals from two colinear ligands. The molecular orbital description of this three-center–four-electron process involves doubly occupied bonding and nonbonding orbitals, but they can be written as well in terms of localized equivalent orbitals.

$$\psi_1 = \psi_H + \lambda\psi_L^1 - \alpha\psi_L^2$$

$$\psi_2 = \psi_H + \lambda\psi_L^2 - \alpha\psi_L^1$$

The essential difference between these bonds is that a single heteroatom orbital is used to form two bonds in complete disregard of simple valence bond ideas. As an example, the structure of SF_4 is shown in the Fig. 1, and this description is in accord with the observed geometries of tetracoordinated sulfur, selenium and tellurium, tri- and pentavalent chlorine, bromine and iodine, and di-, tetra- and hexavalent xenon.

In order to describe the bonding using heteroatoms in their highest valences, i.e., pentacoordinated phosphorus, arsenic and antimony, hexacoordinated sulfur, selenium and tellurium, and heptacoordinated iodine, it is necessary to utilize the doubly occupied s orbitals. Thus hybrid s-p orbitals are utilized to form bonds, some of which resemble hypervalent bonds and some of which resemble covalent bonds. The geometry of these molecules using the heteroatom in its highest valence is basically governed by steric effects and I have called these molecules hypervalent molecules of the second kind, as the distinction between them and those involving p orbitals only, is worthy of notice. I should mention here also that I consider the bonding in transition metal complexes to be essentially of the same form as in these hypervalent molecules, with the empty s and p orbitals of the metal atom combining with the doubly occupied orbitals of the ligand in the usual charge-transfer crystal field assignment of electrons. Thus, when I describe some experiments below, I will include experiments performed with tungsten as the heteroatom.

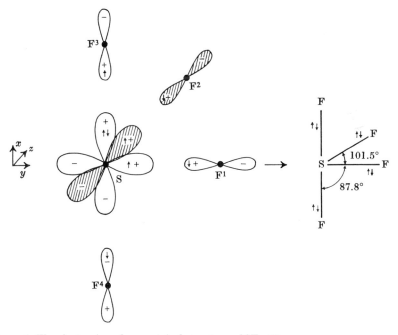

Figure 1 The electronic and geometrical structure of SF_4 (*1*).

Examples among the relatively few known organic molecules using hyper-valent bonds of the first kind are structures I–IV and examples using hyper-valent bonds of the second kind are structures V–VII. The only examples

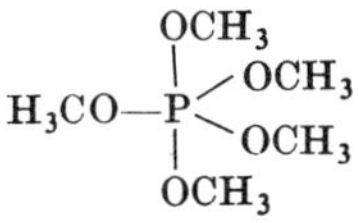

V

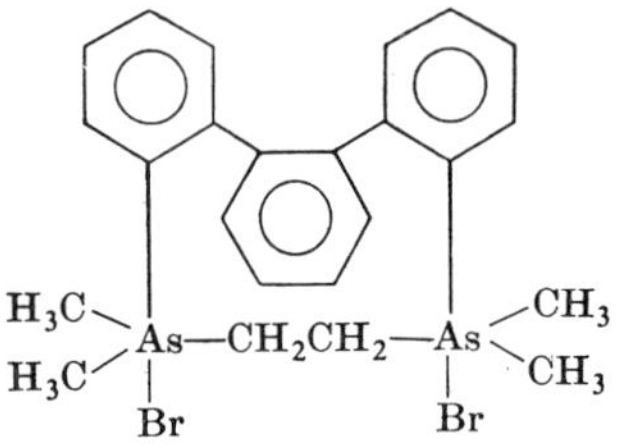

VI

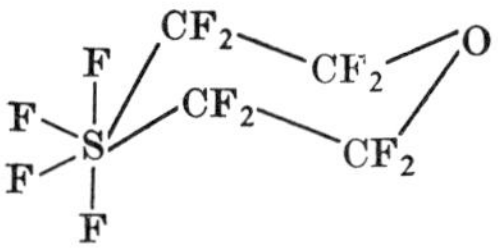

VII

of hypervalent polymers that I am aware of are related to the three structures VIII–X.

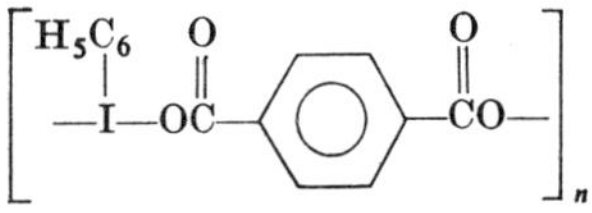

VIII 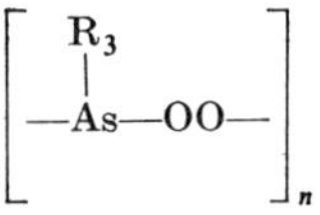   IX 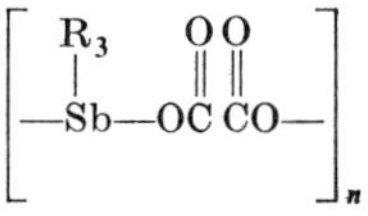   X

One of the most important features of hypervalent molecules is that they can contain more than one type of bond, even to the same monofunctional ligand. As the difference between the types of bonds, which is electronic in nature, is intimately linked to the relative geometries—e.g., hypervalent bonds occur colinearly in pairs—it is not legitimate to consider the rearrangement XI–XII as merely a conformational change, as in cyclohexane. Clearly,

XI   XII

the mechanism involved is a much more drastic one than occurs in conformational changes, where the bonds remain the same and only the second-neighbor interactions are different. As the "inversion" about the phosphorus does not retain configuration and is temperature-dependent, it is best described as a "thermal rearrangement"—at 200°K instead of at the more usual 200°C—rather than as a "conformational inversion." The various possible mechanisms, which must perforce be restricted to "intimate" pair mechanisms, whether classical, nonclassical, solvolytic, etc., can now be investigated following the usual procedures of physical organic chemistry. Also, of course, the mechanisms can be permitted to differ depending on the substituents involved.

There is regrettably little quantitative data available on the rates of unimolecular rearrangements of hypervalent molecules. It appears, however, that with the exception of strained cyclic systems such as I, p-bonded hypervalent molecules do not seem to rearrange easily if at all—recall also the stability to inversion of trivalent phosphorus—whereas hypervalent molecules of the second kind, particularly molecules of phosphorus, arsenic and antimony, do rearrange easily. The plenary lecture by Mislow at this symposium discusses these rearrangements for sulfoxides and phosphene oxides. The subject of rearrangements of octahedrally coordinated hypervalent atoms has not yet been considered due to the paucity of molecules for study.

Rather than discuss the hypothetical temperature-dependence of the hydrocarbon part of organic hypervalent molecules of the first kind such as in XIII, for which rearrangement about the heteroatom is not likely, I will

XIII

(a)   (b)

(c)   (d)

instead briefly examine some studies on phosphorus for which the temperature-dependent behavior has been analyzed extensively in the recent literature.

II. PENTACOORDINATED PHOSPHORUS

The molecule PF_5 is known to have a trigonal bipyramid structure XIV, yet the ^{19}F nuclear magnetic resonance (nmr) spectrum at $-100°C$ does not

$$F-P \begin{matrix} F \\ | \\ | \\ F \end{matrix} \begin{matrix} F \\ \diagup \\ \diagdown \\ F \end{matrix}$$

XIV

distinguish the axial fluorine from the equatorial, which is most reasonably interpreted as implying an activation energy for the rearrangement of less than ~7 kcal/mole. On the other hand, the closely related fluorochlorophosphorane, PF_3Cl_2, which has a room-temperature ^{19}F nmr spectrum similar to that of PF_5, has a low-temperature ^{19}F spectrum showing an AB_2 fluorine system. This has been interpreted as indicative of structure XV, which is

$$F-P \begin{matrix} F \\ | \\ | \\ F \end{matrix} \begin{matrix} Cl \\ \diagup \\ \diagdown \\ Cl \end{matrix}$$

XV

"frozen" for a typical nmr time, of the order of 10^{-1} sec. In fact, the experiment only proves (a) that any rearrangement that takes place at low temperatures must preserve the relative AB_2 configuration, and (b) that there exists at least one activated complex or transition state of energy 7 kcal/mole and at least one stereoisomer differing from XV by an energy of less than this 7 kcal/mole. Incidentally, were the difference in energies of the stereoisomers really small, i.e., $\lesssim 2$ kcal/mole, so that the equilibrium concentration of the second isomer was not totally negligible, then one should be able to detect the nmr spectrum of this stereoisomer superposed on that of XV, or at least should give a lower limit on the energy difference.

To illustrate the configuration retaining nature of an intramolecular rearrangement, consider the phosphorane, $C_6H_5PF_3Cl$, whose temperature-independent ^{19}F nmr spectrum is characteristic of an AB_2 fluorine system, and has been interpreted to indicate the rigid structure XVI. The mechanism

$$F-P \begin{matrix} F \\ | \\ | \\ F \end{matrix} \begin{matrix} Cl \\ \diagup \\ \diagdown \\ C_6H_5 \end{matrix}$$

XVI

most often invoked to explain the rearrangement of pentacoordinated phosphorus is that due to Berry (10), called a "pseudorotation" which takes two equatorial bonds into two axial bonds via a square pyramid transition state in which the bond that remains equatorial in the process is at the apex of the pyramid. Should this mechanism be the operative mechanism, then the reversible rearrangement XVII $\rightleftharpoons$ XIX, in which the phenyl group remains

equatorial, can be said to retain, in part, configuration, since the A fluorine is never exchanged with the B fluorine. The ^{19}F nmr spectrum is therefore consistent not only with structure XVI, but also with the rearrangement XVII $\rightleftharpoons$ XIX, and therefore with a transition state of anywhere less than 15 kcal/mole and a second geometrical isomer of arbitrary energy less than this 15 kcal/mole. The situation is therefore similar to that of methylcyclohexane, where the inversion XX $\rightleftharpoons$ XXI, which is rapid at room temperature,

does not prevent the spectrum from being the broad spectrum similar to that of "frozen" cyclohexane at −60°C. The only way to distinguish between complete rigidity and configuration-retaining rearrangements is by careful temperature-dependent nmr studies, to attempt to observe the effect of the change of equilibrium populations with temperature. Incidentally, I first made this point in 1961 (11), in commenting on the spectra of cis- and trans-decalin observed by Richards and myself in 1958 (12).

The fact that the nmr spectrum of $C_6H_5PF_3Cl$ shows nonequilibrating fluorines at room temperatures, implies that the other two possible stereoisomers, XXII and XXIII, are not attainable, i.e., that the transition state connecting them to XVII or XIX is at an energy $\gtrsim 15$ kcal/mole. Conversely, if any single substituent, such as the phenyl group in $C_6H_5PF_3Cl$, is not permitted to be axial, for whatever reason, the only pseudorotation permitted would be the configuration-retaining rearrangement of XVII $\rightleftharpoons$ XIX. Thus, for example, $C_6H_5PF_2Cl_2$ would exist as two separate products, the rearranging pair XXIV, XXV, and the racemizing dl-pair XXVI, XXVII; and the same would occur for the partially deuterated molecule $CH_3P(OCH_3)_2(OCD_3)_2$.

$$C_6H_5-\overset{\overset{\displaystyle F}{|}}{\underset{\underset{\displaystyle F}{|}}{P}}{\overset{\diagup Cl}{\diagdown Cl}} \quad\rightleftharpoons\quad C_6H_5-\overset{\overset{\displaystyle Cl}{|}}{\underset{\underset{\displaystyle Cl}{|}}{P}}{\overset{\diagup F}{\diagdown F}} \qquad C_6H_5-\overset{\overset{\displaystyle Cl}{|}}{\underset{\underset{\displaystyle F}{|}}{P}}{\overset{\diagup Cl}{\diagdown F}} \quad\rightleftharpoons\quad C_6H_5-\overset{\overset{\displaystyle F}{|}}{\underset{\underset{\displaystyle Cl}{|}}{P}}{\overset{\diagup Cl}{\diagdown F}}$$

$$\text{XXIV} \qquad\qquad \text{XXV} \qquad\qquad \text{XXVI} \qquad\qquad \text{XXVII}$$

The same type of result would, however, obtain if all ligands were permitted to be axial but if the square-pyramidal transition state with two ligands simultaneously in the base were of prohibitively high energy. Were the dimethylamine ligand so restricted, the molecule $[(CH_3)_2N]_2PF_3$ could be separated into two products XXVIII $\rightleftharpoons$ XXIX and XXX, the latter

$$F-\overset{\overset{\displaystyle F}{|}}{\underset{\underset{\displaystyle F}{|}}{P}}{\overset{\diagup N(CH_3)_2}{\diagdown N(CH_3)_2}} \quad\rightleftharpoons\quad F-\overset{\overset{\displaystyle N(CH_3)_2}{|}}{\underset{\underset{\displaystyle F}{|}}{P}}{\overset{\diagup N(CH_3)_2}{\diagdown F}} \qquad F-\overset{\overset{\displaystyle N(CH_3)_2}{|}}{\underset{\underset{\displaystyle N(CH_3)_2}{|}}{P}}{\overset{\diagup F}{\diagdown F}}$$

$$\text{XXVIII} \qquad\qquad \text{XXIX} \qquad\qquad \text{XXX}$$

nonrearranging, so that again an ^{19}F nmr of the AB_2 type would not be sufficient to prove rigidity. There are thus several explanations possible, even within the framework of the all-too-limited pseudorotation theory, to rationalize the AB_2 ^{19}F spectra observed in R_2PF_3 molecules such as H_2PF_3, $\phi ClPF_3$, $\phi[(CH_3)_2N]PF_3$, $C_2H_5[(CH_3)_2N]PF_3$ and $[(C_2H_5)_2N]_2PF_3$.

It is important to notice, however, that we have as yet no evidence to prove that all possible stereoisomers of a pentacoordinated phosphorus molecule are actually metastable structures. In fact, it is likely that in many cases the supposed "stereoisomers" will be of *higher* energy than the associated activated complex or transition state, so that the latter should dissociate into different sets of products rather than rearrange to the stereoisomer, or they will themselves be of higher energy than dissociation products to which they will go directly. Notice also, that whereas the classic Walden inversion about carbon has a trigonal bipyramid as a transition state, phosphorus stereochemistry has a trigonal bipyramid as a stereoisomer, and only under certain circumstances will such a structure be a transition state or an activated complex.

There is some reason to believe that alkyl groups will not, in fact, form hypervalent or axial bonds, as witness the inability to prepare pentamethylphosphorane, tetramethylsulfur, etc., which, of course, does not guarantee (a) their nonexistence, nor (b) the nonexistence of hypervalent methyl groups in other less symmetric systems. (For example, were this to be a general rule, then structure XXXI for CH_3PF_3Cl would be at best a transition state and

XXXI

the ^{19}F nmr spectrum should be of AB_2 type as discussed above for $C_6H_5PF_3Cl$, but at *all* temperatures.) There are, however, several results in the literature of pentavalent phosphorus which serve as either counterexamples to this argument, or as indication of the possible existence of nontrigonal bipyramid structures.

These examples are the following:

(1) The molecule (*13*) H_2PF_3 rearranges intramolecularly above 15°C, which would imply an axial H atom, whereas PH_4F and PH_5 are unknown.

(2) The cyclic molecule $C_4H_8PF_3$ rearranges intramolecularly (*14*) above −70°C, which would imply an axial alkyl group, while the six-membered ring compound $C_5H_{10}PF_3$ is apparently rigid up to 100°C.

(3) The cyclic molecule $(CH_3)_3$-P-OCF$_2$CF$_2$O has been prepared by

Ramirez (*15*) and seems to imply at least one axial methyl group if the ring is closed. (The ^{31}P nmr signal of this molecule is shifted *downfield* by 65 ppm from trimethylphosphine as compared with an *upfield* shift of 209 ppm in going from $(C_2H_5O)_3P$ to $(C_2H_5O)_5P$. While this can be rationalized in several ways, it nevertheless seems quite unusual.)

(4) The oxyphosphoranes XXXII and XXXIII have been observed to

XXXII   XXXIII

isomerize thermally (*15*). Notice that the results could be interpreted equally well as a rearrangement about the equatorial carbon of the four-membered ring to give the two additional isomers which are mirror image of XXXII and XXXIII.

The oxyphosphoranes $\mathbf{XXXIV}$ and $\mathbf{XXXV}$ have been observed to isomerize (*16, 17*) with retention of nonequivalent methoxy groups, and the rearrangement which equilibrates the methoxy groups in $\mathbf{XXXVI}$ has been shown to

$\mathbf{XXXIV}$ $\mathbf{XXXV}$

$\mathbf{XXXVI}$

occur at room temperature although it can be stopped by cooling (*18*). All of these rearrangements require an axial alkyl carbon atom in the pseudo-rotation picture, and notice that this is independent of whether the four- and five-membered rings are permitted to be entirely equatorial or not.

Instead of attempting to resolve these difficulties—which I find decidedly nontrivial—I would rather point out a consequence of the experiments on pentacoordinated phosphorus discussed above which could indicate the likely existence of stable structures about phosphorus other than the trigonal bipyramid. As observed earlier, intramolecular rearrangements about phosphorus occur at room temperatures, which implies the existence of transition "states" which are not trigonal bipyramids within 10–15 kcal/mole of the equilibrium state as well as metastable stereoisomers which are, perhaps, trigonal bipyramids. Since the energy of the pentacoordinated phosphorus bonds, as well as of the trigonal bipyramid, should depend quite drastically on the substituents involved (*1*), it is surely possible that for some sets of ligands one of these different geometrical structures could actually become favored over the trigonal bipyramid, the 10 kcal/mole being easily made up by the different bond energies involved (recall the theoretically arbitrary hybridization which occurs in *sp*-bonded hypervalent molecules). For example, the stable geometrical isomer could in some cases be the square pyramidal structure which is the transition state of the pseudorotation theory; it could also contain equivalent sets of two and three ligands and be in a structure analogous to methylamine or methylnitrate with the carbon–nitrogen distance contracted to zero (notice that the barrier of rotation in this structure might be quite small, and also that such a rotation is also possible for the rearrangement of the trigonal bipyramids of XXXII–XXXVI); and there exists, as well, a myriad of further possibilities. While there is no structural evidence for nontrigonal bipyramidal geometry about phosphorus—although the structure of pentaphenylantimony is a square pyramid while that of pentamethylantimony is unknown—one must retain this as a possibility until rigorously excluded. The nonobservation of such different structures could be invoked as evidence *against* a unimolecular mechanism for these intramolecular rearrangements.

III. SOME PRELIMINARY RESULTS

Before concluding this talk I would like to describe briefly some very preliminary experimental results we have obtained during the past summer in attempts to prepare new organic hypervalent molecules. Last year I prepared what I claimed to be xenon ditrifluoroacetate by reacting xenon difluoride with trifluoroacetic acid, according to the reaction

$$2\,RCOOH + XeF_2 \rightarrow Xe(OCOR)_2 + 2\,HF$$

and I argued that such a reaction was quite general and that one could thereby prepare a series of organic esters, as well as inorganic esters such as xenon dinitrate, and polymers (*19*). Bartlett and co-workers (*20*), using this reaction, have since succeeded in preparing $Xe(OClO_3)_2$ and $Xe(OSO_2F)_2$ as well as $FXe(OClO_3)$ and $FXe(OSO_2F)$, even obtaining an x-ray structure of the latter, thus providing the first solid experimental evidence in favor of my arguments. In order to demonstrate the utility of this reaction, the analogous reaction using the silver salts, and other reactions of organic chemistry, in preparing organic esters of other hypervalent atoms as well as other organic molecules, we have initiated a study of the displacement of fluorine in hypervalent fluorides. We have, in the first series of experiments, reacted PF_5, IF_5 and WF_6 with a variety of organic acids and their salts. All of these fluorides (in excess) react with silver acetate apparently according to the scheme

$$2\,PF_5 + 2\,AgOCOCH_3 \rightarrow Ag^+PF_5OCOCH_3^- \rightarrow PF_4OCOCH_3 + AgF$$

$$2\,IF_5 + AgOCOCH_3 \rightarrow IF_4OCOCH_3 + Ag^+IF_6^-$$

$$WF_6 + AgOCOCH_3 \rightarrow WF_5OCOCH_3 + AgF$$

both with and without solvents. Because of the strong Lewis acid character of IF_5 and PF_5, it is not always easy to separate the product from the silver cation, and there is always the problem of separating the various products of different stoichiometry. When these fluorides are reacted with organic acids, they act in an analogous way, but without the liberation of HF except for WF_6, as

$$PF_5 + RCOOH \rightarrow H^+PF_5OCOR^-$$

$$IF_5 + RCOOH \rightarrow H^+IF_5OCOR^-$$

$$WF_6 + RCOOH \rightarrow WF_5OCOR + HF$$

There are presumably simple conditions under which the HF can be removed in the course of the reaction, but we have not yet had enough time to investigate the various possibilities.

A reaction analogous to these was performed by Couper in 1858, who prepared XXXVII by reacting salicylic acid with phosphorus pentachloride (21)

XXXVII

although the result has been a subject of an important controversy over the years. By a curious accident of history, this molecule figured in one of the most classic, if most neglected, papers of all chemistry, Couper's paper on chemical structure (22), and the cyclic structure about the phosphorus was recognized before that of the much better known compounds, salicylic acid or benzene, itself!

We do not yet know what interest there will be in these organic esters of metalloids and transition metals, other than as prototypes for more complicated molecules, but they are at least likely to be useful as acetylating agents and as joint fluorinating and acetylating agents.

ACKNOWLEDGMENTS

This work is being performed in collaboration with Henry Selig, and I am grateful to the Chemistry Department of the Hebrew University, Jerusalem, for its hospitality. Research was supported in part by the National Science Foundation and the Petroleum Research Fund of the American Chemical Society.

REFERENCES

1.   J.I. Musher, *Angew. Chem.* **81**, 68 (1969); *Angew. Chem. Int. Ed. Engl.* 8, 54 (1969).
2.   G. Wittig, *Bull. Soc. Chim. (France)* **1966**, 1162 .
3.   E.L. Muetterties and R.A. Schunn, *Quart. Rev. (London)* **20**, 245 (1966).

4. R. Schmutzler, *Advan. Fluorine Chem.* **5**, 31 (1965).
5. D. Hellwinkel, *Ber.* **99**, 3628, 3660 (1966).
6. F. Ramirez, *Accounts Chem. Res.* **1**, 168 (1968).
7. F.H. Westheimer, *Accounts Chem. Res.* **1**, 70 (1968).
8. E.L. Muetterties, *J. Am. Chem. Soc.* **90**, 5097 (1968); **91**, 1636 (1969).
9. P.C. Lauterbur and F. Ramirez, *J. Am. Chem. Soc.* **90**, 6722 (1968).
10. R.S. Berry, *J. Chem. Phys.* **32**, 933 (1960).
11. J.I. Musher, *J. Am. Chem. Soc.* **83**, 1146 (1961), footnote 16.
12. J.I. Musher and R.E. Richards, *Proc. Chem. Soc.* **1958**, 230.
13. P.M. Treichel, R.A. Goodrich and S.B. Pierce, *J. Am. Chem. Soc.* **89**, 2017 (1967).
14. E.L. Muetterties, W. Mahler and R. Schmutzler, *Inorg. Chem.* **2**, 613 (1963).
15. F. Ramirez, C.P. Smith and J.F. Pilot, *J. Am. Chem. Soc.* **90**, 6726 (1968).
16. D. Gorenstein and F.H. Westheimer, *Proc. Natl. Acad. Sci. U.S.* **58**, 1747 (1967).
17. F. Ramirez, J.F. Pilot, O.P. Madan and C.P. Smith, *J. Am. Chem. Soc.* **90**, 1275 (1968).
18. D. Gorenstein and F.H. Westheimer, *J. Am. Chem. Soc.* **89**, 2762 (1967).
19. J.I. Musher, *J. Am. Chem. Soc.* **90**, 7371 (1968).
20. N. Bartlett, M. Wechsberg, F.O. Sladky, P.A. Bulliner, G.R. Jones and R.D. Burbank, *Chem. Comm.* **1969**, 703.
21. A.S. Couper, *Compt. Rend.* **46**, 1107 (1858).
22. A.S. Couper, *Compt. Rend.* **46**, 1157 (1858).

Conformational Studies on the Sugar Moieties of Some α-Glycopyranosyl Nucleoside Analogs, Sugar Nucleotides and Related Substances

K. Onodera
Laboratory of Biological Chemistry
Department of Agricultural Chemistry
Kyoto University
Kyoto, Japan

Since the first application of nuclear magnetic resonance (nmr) by Lemieux (*1*) to the investigation of carbohydrates, a number of reports have been published on the conformation of carbohydrates (*2, 3*). The data obtained from nmr spectra are closely related to the conformation of a compound in solution. Therefore, this method has become one of the most powerful tools for the conformational analysis of carbohydrates.

With nmr spectra of carbohydrates, two independent parameters, chemical shifts and coupling constants, are generally used, and it has been found that the signal of an equatorial hydrogen atom appears at lower magnetic field than that of an axial hydrogen atom by about 25 Hz (*4*).

It has also been found with the derivatives of monosaccharides that there is a correlation between orientation and chemical shifts of *O*-acetate methyl signal (τ 7.81–7.85 for axial, τ 7.89–7.98 for equatorial and of *O*-methyl signal (τ6.54–6.58 for axial, τ 6.46–6.47 for equatorial (*2*)). The correlation has successfully been applied to the conformational analysis of monosaccharides as an empirical rule (*5*).

Calculations of free energies (*3*) and Reeves' conformational instability factors (*6*) show that many of the hexopyranoses are more stable in the Cl(D) [or 1C(L)] conformation than in the corresponding alternative one, namely, 1C(D) [or Cl(L)]. However, we found previously (*7*) that (tetra-*O*-acetyl-α-D-mannopyranosyl) theophylline exists in the 1C(D) conformation, in which the aglycon has the equatorial orientation and had been considered to be unfavorable from the viewpoint of Reeves' instability factors (6). Some pyridinium α-D-glycosides (8) and tri-O-acetyl-β-D-xylopyranosyl chloride (9, 10) have also been found to exist in the 1C(D) conformation.

Nucleosides of natural origin generally have the aglycon in *trans* relationship to the hydroxyl group at C-2 of the sugar moiety. Recently, Bhacca and Horton (11) have reported that β-D-ribopyranose tetraacetate exists in a rapid equilibrium between two chair conformations at room temperature. In consideration of these findings, it seemed worthwhile to investigate the conformations of nucleosides, nucleotides and related substances.

The present paper concerns the conformations of the sugar moiety of α-nucleoside analogs of D-mannose, L-rhamnose as well as of nucleotides, sugar nucleotides, and sugar phosphates. D-Mannose and L-rhamnose have the same steric arrangement of groups as lyxopyranose (12), at C-2, C-3 and C-4 of the pyranose ring, and are widely distributed as components of such biologically significant substances as glycosides, polysaccharides, and glycoproteins.

Nuclear magnetic resonance spectra were recorded at 60 MHz with a Varian A-60 spectrometer at its normal operating temperature. Chemical shifts are expressed on the τ scale in parts per million (ppm) downfield displacement from sodium 5,5-dimethylsilapentane-1-sulfonate (DDS) in deuterium oxide, and tetramethylsilane (TMS) in other solvents as internal standards. Coupling constants, in Hz, are the directly measured line spacings. Other experimental details are described in a previous paper (13).

With most of the compounds examined, the signal of the anomeric proton was observed at lower magnetic field than signals of other ring protons. This is due to the combined deshielding effect of the ring oxygen atom and the oxygen or nitrogen atom of the aglycon. The signal of H-1' of 9-α-L-rhamno-pyranosyladenine hydrochloride (I) (Fig. 1) in deuterium oxide appears at τ 3.78 as a sharp doublet. The coupling constant ($J_{1', 2'}$ 4.8 Hz), as estimated from the Karplus equation (14), is smaller than that expected for a *trans* diaxial relationship between H-1' and H-2', and larger than that for a true gauche diequatorial arrangement. The H-2' signal appears at τ 5.00 as a quartet, partly overlapped by the HOD signal. The latter was moved to high magnetic field on heating, to give a well-defined quartet for H-2'. A broad quartet at τ 5.79 was collapsed by irradiating at 46 Hz downfield and the quartet was assigned to H-3'. A two-proton multiplet at high magnetic field is assigned to H-4' and H-5', but was not analyzed because of its complexity.

In the nmr spectrum of I in methyl sulfoxide-d_6 containing a few drops of deuterium oxide, as shown in Fig. 2, the H-1' signal appears at τ 4.00 as a doublet having a spacing of 8.0 Hz, which indicates the *trans*-diaxial orientation of H-1' and H-2'.

A quartet at τ 5.43 is assigned to H-2', and arises through unequal coupling

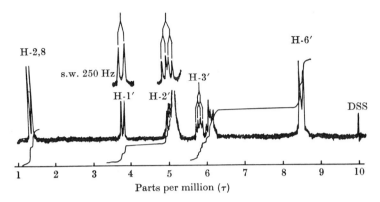

Figure 1 Nuclear magnetic resonance spectrum of 9-(α-L-rhamnopyranosyl)adenine hydrochloride (I) in deuterium oxide.

with H-1′ and with H-3′. The H-3′ and H-4′ signals are observed at τ 6.05 and 6.42, respectively, as triplets. These assignments were confirmed by spin-decoupling experiments. Small values of $J_{2',3'}$ (2.9 Hz), $J_{3',4'}$ (3.6 Hz) and $J_{4',5'}$ (3.5 Hz) are characteristic of vicinal protons having a gauche relationship. In these assignments, several percent of an unfavored conformer might escape detection. In N,N-dimethylformamide as solvent, the H-1′ signal was observed at τ 3.70 as a doublet having $J_{1',2'} = 9.5$ Hz.

In a similar manner, other synthetic α-nucleoside analogs (α-glycosylamine derivatives) and their acetylated derivatives were examined. Tables Ia and Ib show the chemical shifts and first-order coupling constants of the nonacetylated α-nucleoside analogs as well as the presumed conformations, and Tables IIa and IIb show data for their acetylated derivatives. The conformations of these

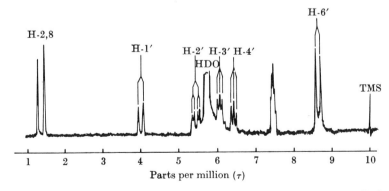

Figure 2 Nuclear magnetic resonance spectrum of 9-(α-L-rhamnopyranosyl)adenine hydrochloride (I) in methylsulfoxide-d$_6$ containing a few drops of deuterium oxide.

Table Ia   Chemical Shifts of α-Glycopyranosides (α-Nucleoside Analogs)

Compound	Solvent	Chemical shifts $(\tau)^a$					
		H-1′	H-2′	H-3′	H-4′	H-5′	H-6′
I	D_2O	3.78d	5.00q	5.79q	6.06	6.15	8.05d
I	Me_2SO-d_6	4.00d	5.43q	6.05t	6.42t	6.10	8.64d
II	D_2O	3.70d	5.39q	5.85q	6.08	6.10	8.53d
II	Me_2SO-d_6	3.82d	5.64q	6.06t			8.64d
III	D_2O	3.78d	5.09q	5.83			
III	Me_2SO-d_6	3.95d	5.58q	6.06t			
IV	D_2O	3.82d	5.46q	5.92q			
IV	Me_2SO-d_6	3.88d	5.62q	6.05			
IV	Pyridine-d_5	3.00d	4.70q	4.9			
V	D_2O	3.99d	5.75				8.50d
V	Me_2SO-d_6	4.10d	5.95				8.64d
VI	D_2O	3.63d	5.07q				8.53d

^{a} Legend: d, doublet; t, triplet; q, quartet.

Table Ib   Coupling Constants of α-Glycopyranosides (α-Nucleoside Analogs)

Compound	Solvent	Coupling constants (Hz)					Conformation presumed
		$J_{1',2'}$	$J_{2',3'}$	$J_{3',4'}$	$J_{4',5'}$	$J_{5',6'}$	
I	D_2O	4.8	3.0	7.0		6.5	C1⇌1C
I	Me_2SO-d_6	8.0	2.9	3.6	3.5	6.8	C1(L)
II	D_2O	6.0	3.3	5.6		6.2	C1⇌1C
II	Me_2SO-d_6	8.2	3.1	3.2		6.7	C1(L)
III	D_2O	6.0	3.0				C1⇌1C
III	Me_2SO-d_6	8.3	3.0	3.2			1C(D)
IV	D_2O	6.2	3.5	5.0			C1⇌1C
IV	Me_2SO-d_6	9.0	3.0				1C(D)
IV	Pyridine-d_5	8.8	3.5				1C(D)
V	D_2O	9.0				6.9	C1(L)
V	Me_2SO-d_6	9.1				7.0	C1(L)
VI	D_2O	8.6	3.5			7.0	C1(L)

acetyl derivatives were also confirmed by observing the chemical shifts of acetate methyl signals in the axial and equatorial orientations, as shown in Tables IIa and IIb.

From the J values of methine and methylene protons as shown in Table Ib, it is found that the pyranoid molecules of 9-α-L-rhamnopyranosyladenine

hydrochloride (I), 7-α-L-rhamnopyranosyltheophylline (II), 9-α-D-manno-
pyranosyladenine hydrochloride (III), 7-α-D-mannopyranosyltheophylline
(IV), and 1-α-L-rhamnopyranosylthymine (V) exist, in methyl sulfoxide-d_6,
predominantly in the so-called alternative conformation (1C for the D series,

Table IIa Chemical Shifts of Acetylated Derivatives of
α-Glycosides (α-Nucleoside Analogs)[a]

Compound	Chemical shifts (τ)[b]					
	H-1'	H-2'	H-3'	H-4'	H-5'	H-6'
VII	3.39d	4.25q	4.53t	5.05t	5.77o	8.43d
VIII[c]	3.55d	4.00q	4.50t	4.94		
VIII[d]		3.39q	3.88t	4.43t		
IX	3.75d	4.70q	4.51t	5.15q	5.70q	8.43d

[a] In chloroform-d.
[b] Legend: d, doublet; t, triplet; q, quartet; o, octet.
[c] In Me_2SO-d_6.
[d] In pyridine-d_5.

Table IIb Coupling Constants of Acetylated Derivatives of
α-Glycopyranosides (α-Nucleoside Analogs)[a]

Compound	Coupling constants (Hz)					Conformation presumed
	$J_{1',2'}$	$J_{2',3'}$	$J_{3',4'}$	$J_{4',5'}$	$J_{5',6'}$	
VII	8.0	3.5	3.5	3.5	7.0	C1(L)
VIII[b]	8.5	3.0	3.5	3.5		1C(D)
VIII[c]	8.3	3.2	3.5	3.5		1C(D)
IX	9.4	3.6	3.6	1.4	7.1	C1(L)

[a] In chloroform-d.
[b] In Me_2SO-d_6.
[c] In pyridine-d_5.

C1 for the L series), in which the aglycons are oriented equatorially. From the
fact that I and IV in N,N-dimethylformamide, and IV in pyridine-d_5, also
exist in this conformation, the remaining α-nucleoside analogs also are probably
very stable in the so-called alternative chair conformation in these solvents.
This conclusion is reasonable, because α-D-mannopyranose and α-L-rhamno-
pyranose (6-deoxy-α-L-mannopyranose) are stereochemical homomorphs
(mirror images), except for the nature of the C-5 substituent.

The nmr spectra of I, II, III and IV, measured in deuterium oxide instead of methyl sulfoxide-d_6, show J values of H-1' that are larger than those expected for the gauche relationship and smaller than those anticipated for the *trans*-axial arrangement of vicinal protons. These J values which were obtained are roughly intermediate between those expected for the C1 and 1C conformations. A possible shift of the aglycon from the true axial orientation in the so-called normal chair conformation is ruled out, because such a distortion would cause the dihedral angle between H-1' and H-2' to increase, giving a J value smaller than that expected for the gauche relationship. Furthermore, consideration of $J_{1',2'}$ (4.8–6.2 Hz), $J_{2',3'}$ (3.0–3.5 Hz) and $J_{1',4'}$ (5.0–7.0 Hz) allows exclusion of the boat forms, because, on the basis of the Karplus equation (*14*), the boat forms possible would have given different J values.

If a pyranoid sugar is in rapid equilibrium between its two chair conformations, with a predominant contribution from the favored one, the J values and chemical shifts observed are determined by the proportion of each conformer present, and are intermediate between those predicted for each separate conformer (*8, 9*). The spectrum of IV, measured at 4°C, gives signals essentially the same as those measured at room temperature. Much lower temperatures would be required for nmr spectral observation of separate conformers.

From J values observed with I, II, III and IV in deuterium oxide, and with the use of $J_{1',2'} = 1.7$ Hz and 8.5 Hz for diequatorial (*15*) and diaxial orientations, respectively, for α-D-mannopyranose as a reference standard, the proportion of the C1 and 1C conformations, and free-energy differences ($\Delta G°$) between the two conformations can be estimated approximately (Table III). The possible influence of the aglycon and solvent on the J values cannot be excluded.

On the basis of these results, it is reasonable to conclude that I, II, III, IV and V exist in methyl sulfoxide, N,N-dimethylformamide and pyridine, and V and VI in deuterium oxide, predominantly in the so-called alternative chair conformation, and that I, II, III and IV exist in rapid conformational equilibrium in deuterium oxide at room temperature.

With regard to the acetyl derivatives of these compounds, the alternative conformation is favored, with various solvents, by 7-(2,3,4-tri-O-acetyl-α-L-rhamnopyranosyl)theophylline (VII), 7-(2,3,4,6-tetra-O-acetyl-α-D-mannopyranosyl)theophylline (VIII) and 1-(2,3,4-tri-O-acetyl-α-L-rhamnopyranosyl)thymine (IX). It must be recognized that a small percentage of the unfavored conformer may escape detection (*11*).

The fusion procedure is one of the most widely used reactions for the synthesis of nucleosides, and a number of nucleosides and their analogs have been synthesized by this procedure.

In the course of a study on fusion procedure, an abnormal reaction has been found in our laboratory that leads to the production of unsaturated nucleoside analogs (*16*). This reaction is of interest from the chemical and biochemical points of view, with respect to the presence of unsaturation in the sugar moiety of some nucleoside antibiotics (*17*).

We have synthesized 7-(2,4-di-*O*-acetyl-3,6-dideoxy-α-L-*erythro*-hex-2-enopyranosyl)theophylline (*16*) (X), 7-(2,4,6-tri-*O*-acetyl-3-deoxy-α-D-*erythro*-hex-2-enopyranosyl)theophylline (XI), and 7-(2,4-di-*O*-acetyl-3-deoxy-α-D-*glycero*-pent-2-enopyranosyl)-theophylline (XII) together with three β-nucleosides analogs of such kind. The syntheses of these compounds will be reported elsewhere.

It has been reported that D-glucal triacetate (*18*) is expected to adopt a half-chair conformation, and the unusual chemical shifts of the H-1 and H-2

Table III   Calculated and Observed *J* Values and Free-Energy Differences

Compound	Proportion of conformers (%)		Calculated *J* values (Hz)			Observed *J* values (Hz)			$-\Delta G^\circ$ (kcal/mole)[a]
	C1	1C	$J_{1',2'}$	$J_{2',3'}$	$J_{3',4'}$	$J_{1',2'}$	$J_{2',3'}$	$J_{3',4'}$	
I (L)	45	55	4.8	3.3[b]	6.6	4.8	3.3	7.0	+0.1
II (L)	60	40	5.8	3.3	5.4	6.0	3.3	5.6	−0.3
III (D)	40	60	5.8	3.3	5.4	6.0	3.0		−0.3
IV (D)	35	65	6.1	3.3	5.1	6.2	3.5	5.0	−0.4

[a] Calculated from the equation, $-\Delta G^\circ = RT \ln K$, where $K = N_N/N_A$ (N_N = molar proportion of the normal [C1(D)] and N_A alternative [1C(D)] conformation or 1C(L) and C1(L), respectively.

[b] This is an average value of $J_{2',3'}$ observed, and is used as a reference for the true gauche relationship of the vicinal protons.

protons of the compound are considered to be due to cyclic α,β unsaturation. However, little work has been reported on the nmr spectrum of the unsaturated pyranoid ring.

Comparison of the nmr spectrum (CDCl$_3$) of the compound X with that of 7-(2,3,4-tri-*O*-acetyl-α-L-rhamnopyranosyl)theophylline (*13*), showed that the ring-proton signal had disappeared and that two acetate methyl signals appeared (Fig. 3). From the H-1′ singlet and the H-3′ quartet ($J_{3',4'}$, 3.0 Hz), the double bond is assigned the position C-2–C-3. The acetoxy group is considered to be attached to C-2, in comparison with the data obtained with 1,2,3-tri-*O*-acetyl-3,6-dideoxy-α-D-*erythro*-hex-2-enopyranose (*19*). The values $J_{4',5'} = 8.0$ Hz, $J_{3',4'} = 3.0$ Hz, and the equatorial orientation (*20*) at Me-C-5′ (τ 8.73) indicate that the compound is a 1 H type in a half-chair conformation with aglycon in a quasi-axial orientation.

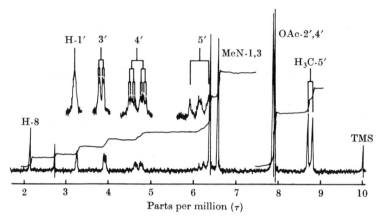

Figure 3 Nuclear magnetic resonance spectrum of 7-(2,4-di-*O*-acetyl-3,6-dideoxy-α-L-*erythro*-hex-2-enopyranosyl)theophylline (X) in CDCl₃.

In a similar way, two other unsaturated α-nucleoside analogs were examined with their acetylated derivatives. Tables IVa and IVb show chemical shifts and first-order coupling constants of the compounds. The assignments were confirmed by spin-decoupling experiments. It is suggested that the compounds XI and XII adopt the same conformations as that of the compound X. However, further investigations are desirable as to the conformation of unsaturated sugar molecules.

Purine and pyrimidine 5'-(α-D-pyranosyl pyrophosphates) ("sugar nucleotides") are well-known as glycosyl donors in the biosynthesis of numerous glycosides, including oligo- and polysaccharides. In order to elucidate the conformational changes during the glycosyl transfer to polysaccharides, it

Table IVa Chemical Shifts of Acetylated α-Glycalpyranosides (Unsaturated α-Nucleoside Analogs)[a]

Compound	Chemical shifts (τ)[b]					OAc[c]		
	H-1'	H-3'	H-4'	H-5'	H-6'			
X	3.23	3.90q	4.69o	6.13o	3.73d	7.86 (3)	7.91 (3)	
XI	3.17	3.84q	4.39o	6.13o	5.80	7.83 (3)	7.85 (3)	7.96(3)
XII	3.18d	3.80t	4.61o	6.18[d]		7.87 (3)	7.89 (3)	

[a] In chloroform-d.
[b] Legend: d, doublet; t, triplet; q, quartet; o, octet.
[c] Three protons.
[d] Two protons.

Table IVb Coupling Constants of Acetylated α-Glycalpyranosides (Unsaturated α-Nucleoside Analogs)[a]

Compound	Coupling constants (Hz)						
	$J_{1',3'}$	$J_{1',4'}$	$J_{1',5'}$	$J_{3',4'}$	$J_{3',5'}$	$J_{4',5'}$	$J_{5',6'}$
X	1.0	2.0	0.5	3.0		8.0	6.0
XI	1.0	1.5		2.5		8.0	4.5
XII	1.0			6.0	5.5	$2.5,^b\ 2.0^c$	

[a] In chloroform-d.
[b] $J_{4',5'a}$.
[c] $J_{4',5'b}$.

is of significance to examine the conformation of pyranose moieties in the molecule of sugar nucleotides, nucleotides and sugar phosphates.

The presence of C1 conformation is found in α-D-mannopyranose moieties of α-D-mannopyranose-1-phosphate (XIII) and guanosine 5'-(α-D-mannopyranosyl pyrophosphate) (XIV) and in 2-acetamido-2-deoxy-D-hexopyranose moiety of uridine 5'-(2-acetamido-2-deoxy-α-D-glucopyranosyl pyrophosphate) (XV).

As shown in Table V, H-1 signals of α-D-hexopyranose-1-phosphates appear at τ 4.32–4.66 as quartet due to the spin–spin couplings of H-1 with both H-2 and P. In the nmr spectrum of XIII, H-1 signal appears at τ 4.66 as quartet with $J_{1,2} = 1.5$ Hz and $J_{1,P} = 8.5$ Hz. The α-D-mannopyranose moiety involves two axial C–H bonds at C-1 and C-2 in 1C conformation, and the projected angle between both the C–H bonds is estimated at ~180°, according to the Karplus equation (15). Therefore, the small value (~1.0 Hz) of $J_{1,2}$ does not support 1C conformation but strongly supports C1 conformation for XIII. The nmr spectrum of acetylated derivatives of XIII reveals the presence of O-acetate methyl signals in one axial and three equatorial orientations in the pyranose moiety. The observation also supports C1 conformation for XIII.

In D-ribofuranose moieties of nucleotides and sugar nucleotides, H-1' signals appear at τ 4.02–4.09 as doublet with $J_{1',2'} = 4.0$–5.0 Hz. In α-D-hexopyranose moieties of sugar nucleotides, H-1' signal appears at τ 4.32–4.66 as quartet with $J_{1',2'} = 1.0$–3.0 Hz and $J_{1',P} = 7.0$–8.5 Hz. The difference of chemical shifts makes it possible to assign H-1' signals of both monosaccharide moieties, D-mannose and D-ribose for the conformational analysis on the basis of the spin–spin coupling constants of H-1' and H-2'. The nmr spectrum of XIV is shown in Fig. 4. In D-mannopyranose moiety, H-1' signal appears at τ 4.39 as quartet with $J_{1',2'} = $ ~1.0 Hz and $J_{1',P} = 7.5$ Hz, and H-1' signal of D-ribofuranose moiety appears at τ 4.10 as doublet with $J_{1',2'} = 5.0$ Hz.

Table V H-1' Signals of D-Hexopyranose and D-Ribofuranose Moieties in the Molecules of Some α-D-Hexopyranose-1-phosphates, Sugar Nucleotides, and Nucleotides

Compound	Solvent	D-Hexopyranose			D-Ribofuranose	
		Chemical shifts (τ) H-1'	Coupling constants (Hz) $J_{1',2'}$	$J_{1',P}$	Chemical shifts (τ) H-1'	Coupling constants (Hz) $J_{1',2'}$
α-D-Mannopyranose-1-phosphate[a]	D$_2$O	4.66q	1.5	8.5		
α-D-Glucopyranose-1-phosphate[b]	D$_2$O	4.60q	3.0	7.0		
α-D-Galactopyranose-1-phosphate[c]	D$_2$O	4.32q	1.0	7.0		
Guanosine 5'-(α-D-mannopyranosylpyrophosphate)[d]	D$_2$O	4.39q	1.0	7.5	4.10d	5.0
Uridine 5'-(α-D-glucopyranosylpyrophosphate)[e]	D$_2$O	4.39q	3.0	7.0	4.02d	4.0
Uridine 5'-(2-acetamido-2-deoxy-α-D-glucopyranosylpyrophosphate)[c]	D$_2$O	4.38q	3.0	7.0	4.02d	4.0
Uridine 5'-(α-D-glucopyranosyluronic acid pyrophosphate)[c]	D$_2$O	4.38q	3.0	7.0	4.02d	4.0
Guanosine 5'-monophosphate[c]	D$_2$O				4.09d	5.0
Uridine 5'-monophosphate[c]	D$_2$O				4.02d	4.0

[a] Cyclohexylammonium salt.
[b] Potassium salt.
[c] Sodium salt.
[d] Barium salt.
[e] Lithium salt.

The small value (~1.0 Hz) of $J_{1',2'}$ supports C1 conformation for α-D-manno-pyranose moiety of XIV. In spite of the relatively large aglycon, the conformational inversion is not found in α-D-mannopyranose moiety of XIV.

With regard to XV, the examination of chemical shifts of N-acetate methyl signals in axial and equatorial orientations, with the use of a number of N-acetyl derivatives of D-hexopyranoses, indicates the presence of N-acetyl group in equatorial orientation, which supports C1 conformation for XV.

As mentioned above, Lemieux and Morgan (8) reported a quaternized nitrogen atom at the anomeric position in axial orientation to be a factor for

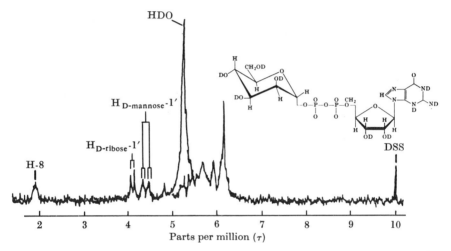

Figure 4   Nuclear magnetic resonance spectrum of guanosine 5'-(α-D-mannopyranosyl-pyrophosphate) recorded at 60 MHz in D_2O.

the conformational inversion. The results presented in this paper indicate that a tertiary nitrogen atom in axial orientation causes conformational inversion or conformational equilibration. The hexopyranose moieties of sugar phosphates and sugar nucleotides (16) as well as O-α-glycopyranosides (13) were found to exist in the normal conformation. The C-1–N bond distance in α-glycosylamines (α-nucleoside analogs) is shorter than the C-1–O bond in α-glycosides. Therefore, the aglycons in α-glycosides may have little influence on the conformation from the viewpoint of steric, mesomeric, and polar effects, but the C-1 substituent in α-nucleoside analogs may have a strong influence on the conformation, and may bring about conformational equilibrium or inversion. The fact that the alternative conformation is favored in spite of many destabilizing factors indicates that the axial aglycon greatly destabilizes the normal conformation, probably by the reverse anomeric effect (8) and by steric hindrance between syn-axial hydrogen atoms and the axial aglycons.

The effect of solvents on the shape of the pyranose ring is clearly shown in this study. Methyl sulfoxide-d_6, pyridine-d_5, and N,N-dimethylformamide are considered to solvate strongly the axial hydroxyl groups of pyranose derivatives, and may stabilize axial substituents sufficiently to give the alternative conformation.

REFERENCES

1. R.U. Lemieux, R.K. Kullnig, H.J. Berstein and W.G. Schneider, *J. Am. Chem. Soc.* **80**, 6098 (1958).
2. L.D. Hall, *Advan. Carbohydrate Chem.* **19**, 23 (1964).
3. E.L. Eliel, N.L. Allinger, S.J. Angyal and G.A. Morrison, "Conformational Analysis," pp. 351–432, Wiley (Interscience), New York, 1965.
4. E.L. Eliel, M.H. Gianni, T.H. Williams and J.B. Stothers, *Tetrahedron Letters* **1962**, 741.
5. K. Onodera and S. Hirano, *Biochem. Biophys. Res. Commun.* **25**, 239 (1966).
6. R. E. Reeves, *J. Am. Chem. Soc.* **71**, 215 (1949); *Advan. Carbohydrate Chem.* **6**, 107 (1951).
7. K. Onodera, S. Hirano, F. Masuda and N. Kashimura, *J. Org. Chem.* **31**, 2403 (1966).
8. R.U. Lemieux and A.R. Morgan, *Can. J. Chem.* **43**, 2205 (1965).
9. C.V. Holland, D. Horton and J.S. Jewell, *J. Org. Chem.* **32**, 1818 (1967).
10. L.D. Hall and J.F. Manville, *Carbohydrate Res.* **4**, 512 (1967).
11. N.S. Bhacca and D. Horton, *J. Am. Chem. Soc.* **89**, 5993 (1967).
12. M. Rudrum and D.F. Shaw, *J. Chem. Soc.* **1966**, 52.
13. K. Onodera, S. Hirano and F. Masuda, *Carbohydrate Res.* **7**, 27 (1968).
14. M. Karplus, *J. Chem. Phys.* **30**, 11 (1959); *J. Am. Chem. Soc.* **85**, 2870 (1963).
15. R.U. Lemieux and J.D. Stevens, *Can. J. Chem.* **44**, 249 (1966).
16. K. Onodera, S. Hirano, F. Masuda and T. Yajima, *Chem. Commun.* **23**, 1538 (1968).
17. J.J. Fox, K.A. Watanabe and A. Bloch, *Progr. Nucleic Acid Res. Mol. Biol.* **5**, 251 (1965).
18. L.D. Hall and L.F. Johnson, *Tetrahedron* **20**, 883 (1964).
19. R.J. Ferrier and G.H. Sankey, *J. Chem. Soc. (C)* **1966**, 2345.
20. A.C. Richardson and K.A. McLauchlan, *J. Chem. Soc.* **1962**, 2499.

Conformational Equilibria in Low- and High-Molecular-Weight Paraffins

S. Pucci
Centro di Chimica delle Macromolecole
Pisa, Italy

M. Aglietto
Istituto di Chimica Organica Industriale dell'Università
Pisa, Italy

P.L. Luisi and P. Pino
Technisch-Chemisches Laboratorium
Eidgenössische Technische Hochschule
Zurich, Switzerland

While investigating the relationship between optical activity and conformation in the isotactic polymers of optically active α-olefins (I–IV), it became apparent that when the asymmetric carbon atoms present in the lateral chains were in α or β positions with respect to the main chain (I–III), the molar optical activity at the D line referred to one monomeric unit was, in absolute value, about 10 times higher than that of the low-molecular-weight paraffins having similar structure (V–VIII). For I–VIII the values of $[\varPhi]_D^{25}$ are the following: I, 161 (*1*); II, −130.0 (*6*); III, 288 (*1*); IV, 68.1 (*1*); V, −11.4 (*2*); VI, −14.4 (*3*); VII, 21.3 (*4*); VIII, 13.1 (*5*). The enhancement was about 5 times when the asymmetric carbon atom of the lateral chain was in the γ position with respect to the principal chain (IV and VIII), and no enhancement was observed when the asymmetric carbon atom was in the δ position. In the last case, the stereoregularity of the polymer could not be determined (*1*).

As the electronic transition at the longest wavelength responsible for the rotation (evaluated from a one-term Drude equation) occurs at about the same wavelength in low- and high-molecular-weight hydrocarbons (*6*), the optical rotation enhancement observed in the polymers was attributed to a different

203

position of the conformational equilibria in the two cases (6–8). According to the above hypothesis, particular conformations of short, saturated hydrocarbon chains (containing 5–7 carbon atoms) should exist, which possess a molar optical rotation at 589 mμ above 100°.

The existence of this kind of conformation having a high rotation was already foreseen in a completely different way by Brewster (9). In order to predict the optical rotation of paraffins, he admits that the relatively low value of the optical rotation found in the most common cases arises from the coexistence of all the "allowed" conformations, some of which have remarkably high optical rotation.

$$
\begin{array}{cccc}
\sim\!\!\!-CH\!-\!CH_2\!-\!\!\!\sim & H_3C\!-\!CH\!-\!CH_3 & \sim\!\!\!-CH\!-\!CH_2\!-\!\!\!\sim & H\!-\!CH\!-\!CH_3 \\
\quad | & \quad | & \quad | & \quad | \\
H_3C\!-\!C\!-\!H & H_3C\!-\!C\!-\!H & H\!-\!C\!-\!CH_3 & H_3C\!-\!C\!-\!H \\
\quad | & \quad | & \quad | & \quad | \\
C_2H_5 & C_2H_5 & (CH_2)_3 & (CH_2)_3 \\
\mathbf{I} & \mathbf{V} & \quad | & \quad | \\
 & & CH\!-\!CH_3 & CH\!-\!CH_3 \\
 & & \quad | & \quad | \\
 & & CH_3 & CH_3 \\
 & & \mathbf{II} & \mathbf{VI}
\end{array}
$$

$$
\begin{array}{cccc}
\sim\!\!\!-CH\!-\!CH_2\!-\!\!\!\sim & H_3C\!-\!CH\!-\!CH_3 & \sim\!\!\!-CH\!-\!CH_2\!-\!\!\!\sim & H_3C\!-\!CH\!-\!CH_3 \\
\quad | & \quad | & \quad | & \quad | \\
CH_2 & CH_2 & (CH_2)_2 & (CH_2)_2 \\
\quad | & \quad | & \quad | & \quad | \\
H_3C\!-\!C\!-\!H & H_3C\!-\!C\!-\!H & H_3C\!-\!C\!-\!H & H_3C\!-\!C\!-\!H \\
\quad | & \quad | & \quad | & \quad | \\
C_2H_5 & C_2H_5 & C_2H_5 & C_2H_5 \\
\mathbf{III} & \mathbf{VII} & \mathbf{IV} & \mathbf{VIII}
\end{array}
$$

For saturated hydrocarbon chains of 5–7 carbon atoms, the highest predicted values according to Brewster (9) are in the range of that found for the optically active isotactic poly-α-olefins.

In order to confirm the validity of our explanation of the high optical activity found in isotactic poly-α-olefins as well as the soundness of the Brewster approach to the calculation of optical activity, we have synthesized some optically active low-molecular-weight paraffins having only a few "allowed" conformations. In the present paper, we discuss the results we have obtained, as well as some aspects of the optical activity in isotactic poly-α-olefins.

I. SYNTHESIS OF LOW-MOLECULAR-WEIGHT MODELS

In order to predict the structure of low-molecular-weight paraffins resembling the monomeric units of (I–IV) and having optical activity in the range of that found for the monomeric units of the optically active poly-α-olefins, we needed both a method to evaluate the optical activity of particular conformations and

Table I Comparison between the Maximum Experimental Value of Molar Rotatory Power of Some Optically Active Paraffins and the Value Calculated According to Brewster (9)

Optically active paraffin	[Φ]$_D^{25}$		Molar rotatory power of the allowed conformations having the highest rotation	Number of allowed conformations
	Maximum experimental value	Value calculated according to Brewster (9)		
(S)-2,3-Dimethylpentane	−11.4 (2, 9a)[a]	−15.0	±120	4
(S)-3-Methylhexane	+ 9.9 (4)	+10.0	±180	6
(S)-2,4-Dimethylhexane	+21.3 (4)	+20.0	±180	3
	+22.9 (5)			
(S)-3-Methylheptane	+11.4 (4)	+12.0	±240	15
(S)-2,5-Dimethylheptane	+11.7 (9b)	+13.3	±240	9
	+13.1 (5)			
(S)-3-Methyloctane	+13.3 (9c)	+13.3	±300	36
(S)-2,6-Dimethyloctane	+14.4 (3)	+14.3	±300	21
(4R,6R)-2,4,6,8-Tetramethylnonane	−23.0 (9d)	−24.0	±360	5
(4S,8S)-2,4,6,8,10-Pentamethylundecane	+31.4 (9e)	+34.3	±480	7

[a] At 20°C.

Table II Conformational Analysis and Optical Activity of Model Compounds of Poly-(S)-4-methyl-1-hexene

Model compound	Newman projections of allowed[a] conformations					$[\Phi]_D$ Calculated according to Brewster[a]	$[\Phi]_D$ Average according to Brewster[a]
	(a)	(b)	(c)	(d)	(e)		
$H_3C-\underset{(a)}{CH}-CH_3$ $\quad\quad|_{(b)}CH_2$ $H_3C-\underset{(c)}{C}-H$ $\quad\quad|$ $\quad\quad C_2H_5$ **IX**	Newman projection (CH_3, H, CH_3, C_4H_9, H, H)	Newman projection (C_3H_7, CH_3, H, C_2H_5, H, H)	Newman projection (C_4H_9, H, H_3C, CH_3, H, H)			+60	+20
	Newman projection (CH_3, H, CH_3, C_4H_9, H, H)	Newman projection (C_3H_7, CH_3, H, C_2H_5, H, H)	Newman projection (C_4H_9, CH_3, H, H_3C, H, H)			+180	
	Newman projection (CH_3, H, CH_3, H_9C_4, H, H)	Newman projection (C_3H_7, H, H_5C_2, CH_3, H, H)	Newman projection (C_4H_9, H, H_3C, H, H, H)			−180	
$H_5C_2-\underset{(b)}{CH}-\underset{(a)}{C_2H_5}$ $\quad\quad\quad|_{(c)}CH_2$ $H_3C-\underset{(d)}{C}-H$ $\quad\quad\quad|_{(e)}$ $\quad\quad\quad C_2H_5$ **X**	Newman projection (CH_3, C_5H_{11}, H, C_2H_5, H, H)	Newman projection (CH_3, C_2H_5, H, C_5H_{11}, H, H)	Newman projection (C_2H_5, C_4H_9, H, H_5C_2, H, H)	Newman projection (C_5H_{11}, CH_3, H, C_2H_5, H, H)	Newman projection (C_6H_{13}, CH_3, H_3C, H, H, H)	+300	+60
	Newman projection (CH_3, C_5H_{11}, H, C_2H_5, H, H)	Newman projection (CH_3, C_2H_5, H, C_5H_{11}, H, H)	Newman projection (C_2H_5, C_4H_9, H, H_5C_2, H, H)	Newman projection (C_5H_{11}, CH_3, H, C_2H_5, H, H)	Newman projection (C_6H_{13}, CH_3, H_3C, H, H, CH_3)	+180	

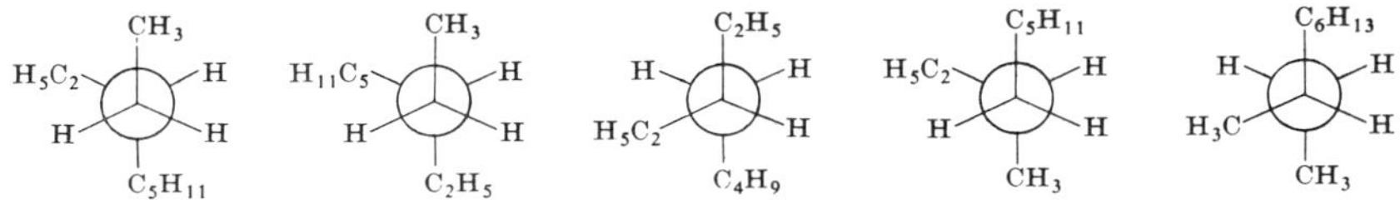

-300

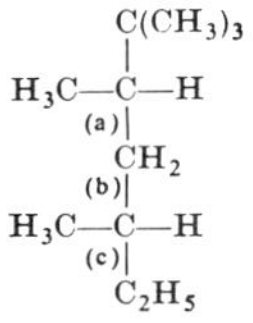

-180

$+180$

$+60$

-180

$+120$

$\begin{array}{c}C(CH_3)_3\\ |\\ (a)\\ H_3C-C-H\\ |\\ (b) CH_2\\ |\\ (c)\\ H_3C-C-H\\ |\\ C_2H_5\end{array}$

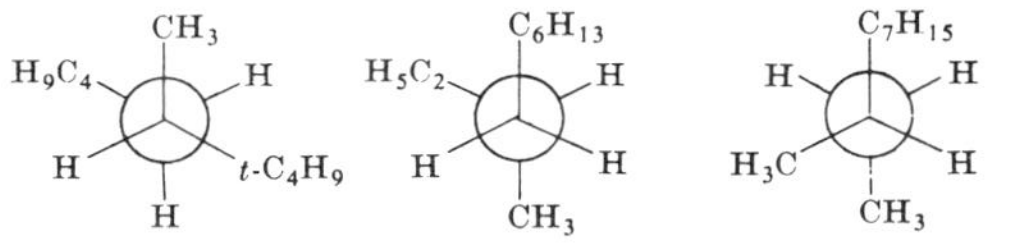

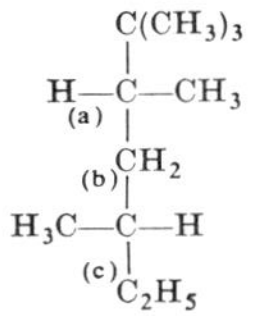

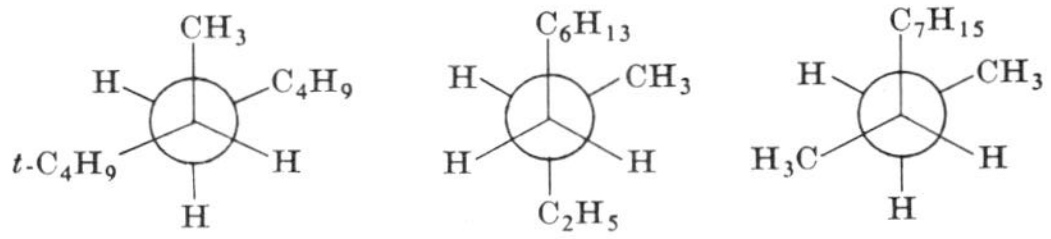

XIa

$\begin{array}{c}C(CH_3)_3\\ |\\ (a)\\ H-C-CH_3\\ |\\ (b) CH_2\\ |\\ (c)\\ H_3C-C-H\\ |\\ C_2H_5\end{array}$

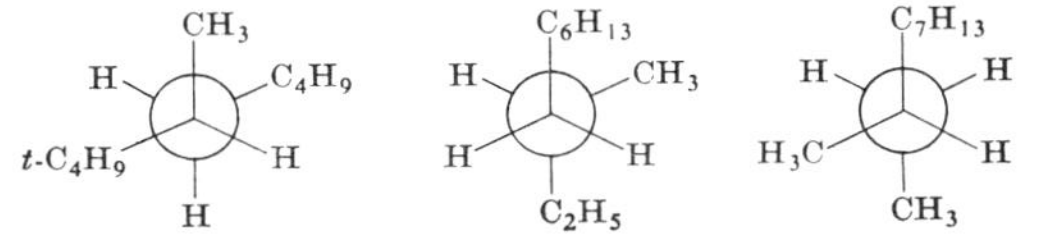

XIb

(continued)

Table II—Continued

Model compound	Newman projections of allowed[a] conformations					$[\Phi]_D$ Calculated according to Brewster[a]	$[\Phi]_D$ Average according to Brewster[a]
	(a)	(b)	(c)	(d)	(e)		
(a) CH(CH$_3$)$_2$ (b) H—C—CH$_3$ (c) CH$_2$ (e) H$_5$C$_2$—C—H (d) C$_2$H$_5$ **XIII**						+300 +180 +240 −300 −240	+105
(a) C(CH$_3$)$_3$ (b) H$_3$C—C—H (c) CH$_2$ (e) H$_5$C$_2$—C—H (d) C$_2$H$_5$ **XII**						−240	−240

[a] J. H. Brewster, *J. Am. Chem. Soc.* **81**, 5475 (1959).

a suitable method of conformational analysis in order to estimate roughly the position of the conformational equilibrium.

Since the Brewster approach (9) has given excellent results for all the optically active paraffins which were known before the present research (Table I), we have used the Brewster rules to predict structures similar to I–IV, and in particular to III, for which only few conformations were allowed having high optical activity.

In Table II, the results of a preliminary conformational analysis and evaluation of $[\Phi]_D$ according to Brewster are reported for some paraffins resembling the monomeric unit of III; among these, only IX was known in literature; X–XIII were synthesized as shown in Schemes I and II through steps which do not involve racemization or inversion of configuration. The two diastereoisomers of XI were separated by crystallization from liquid propane and the absolute configuration of the asymmetric carbon atom in position 3 was assigned, as already described (10). Experimental details concerning the synthetic work have been published (18).

II. OPTICAL ROTATION OF "CONFORMATIONALLY RIGID"* PARAFFINS

The minimum values of the maximum rotatory power of compounds IX–XIII are reported in Table III together with the optical rotation temperature coefficients. The sign of the rotation found for the synthesized paraffins corresponds to that foreseen by the Brewster method. On the contrary, the magnitude of the experimental value of optical rotation corresponds to the one calculated, only in the case of IX and of XIb, the experimental value being less than 15% higher than the calculated values (6, 10).

In all the other cases, the experimental values are much smaller than the calculated ones; the above discrepancies might be due to the rough approximations existing in the Brewster calculation of $[\Phi]_D$ and particularly to the assumptions (a) of the existence of staggered conformations only, (b) of the equal statistical weight for all the allowed conformations and (c) of the statistical weight zero attributed to the forbidden conformations.

An indication that at least in the case of XIa the conformational analysis indicating a very large prevalence of a single conformation is substantially correct, is given by the comparison of the infrared (ir) spectra in the liquid, glass and crystalline states (12). Contrary to what happens in general (when several conformations are allowed in the liquid state and only one in the solid state), the ir spectra of XIa in the liquid state are simpler than those in the

* We use the expression "conformationally rigid paraffins"(6, 10, 11), although it is not a particularly suitable one, to indicate paraffins for which only one or few conformations are "allowed" among the many which are possible and participate in the conformational equilibrium.

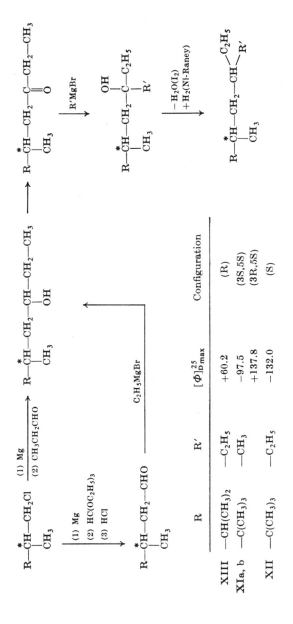

Scheme I

solid state, and very similar. This fact can be taken as an indication of the presence of substantially the same single conformation in the liquid and solid states, the spectrum in the solid state being more complicated because of the presence of bands arising from intermolecular interactions in the crystals.

Some attempts to improve the agreement between calculated and experimental value, attributing to all the allowed conformations their statistical weight deduced from the known literature data for the energy difference between *trans* and gauche conformations, gave very unsatisfactory results.

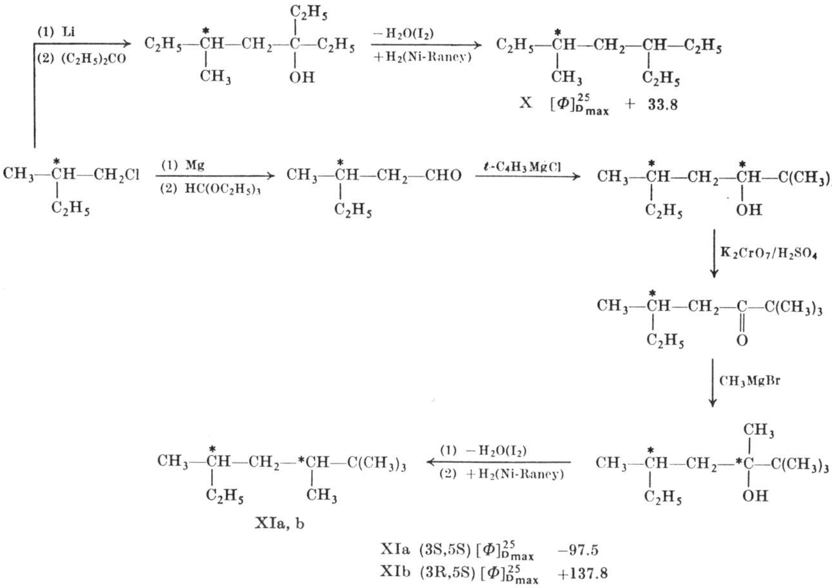

Scheme II

A better agreement between experimental and calculated values has been achieved considering as "allowed," the conformations in which, in the Newman projections, three substituents different from hydrogen are in the vicinal position and which are considered "forbidden" according to the Brewster rules (see Table III).

In any case, the poor agreement between experimental and calculated values does not seem to be attributable to the presence of t-butyl groups, as it has been found also in compounds X and XIII and not in XIb.

Comparing the calculated and experimental values of IX, where a very good agreement exists, with X, XII and XIII, it seems that the failure of Brewster calculations is connected with the presence of a 3-pentyl group; however, this

Table III Maximum Experimental and Calculated $(9)^{a,b}$ Values of Molar Rotatory Power and Rotation Temperature Coefficients of Some Optically Active Paraffins

Optically active paraffin	$[\Phi]_D^{25}$ Maximum experimental value	Calculated values		Number of allowed conformations		$\dfrac{\Delta[\Phi]_D^{\;c}}{\Delta T}$
		a	b	a	b	
IX (S)-2,4-Dimethylhexane	+ 22.9 (5)	+ 20	+ 12	3	5	-0.065
X (S)-3-Ethyl-5-methylheptane	+ 33.8	+ 60	+ 30	3	14	+0.152
XIa (3S,5S)-2,2,3,5-Tetramethylheptane	- 97.5	-180	-150	1	2	-0.128
XIb (3R,5S)-2,2,3,5-Tetramethylheptane	+137.8	+120	+120	2	3	
XII (S)-5-Ethyl-2,2,3-trimethylheptane	-132.0	-240	-165	1	4	+0.155
XIII (R)-5-Ethyl-2,3-dimethylheptane	+ 60.2	+105	+ 67.5	4	16	-0.056

[a] According to Brewster (9).

[b] According to Brewster (9), but considering as allowed also the conformations having one double vicinal gauche conformation, i.e., three consecutive neighbor substituents different from H in the Newman projection.

[c] Between 25 and 110°C.

cannot be the only reason, as a discrepancy of the same order of magnitude has been found in XIa.

The existence of a rather large temperature coefficient of the rotation in XIa, XIb and XII has been interpreted considering that the statistical weight of conformations of higher energy having a high positive rotation increases by increasing the temperature.

The order of magnitude of $\Delta[\Phi]_D/\Delta T$ that was found corresponds to the increase of the expected contribution of conformations having one 1–5 carbon-to-carbon nonbonded high-energy interaction or two vicinal gauche conformations if that interaction would cause an average increase of about 2500–3000 cal with respect to the allowed conformations.

For instance, in the case of (3S,5S)-2,2,3,5-tetramethylheptane, which has only one allowed conformation, we have ten forbidden conformations having only one high-energy strain (1–5 carbon-to-carbon nonbonded interactions or two vicinal gauche conformations). Calculating the molar optical rotation for these conformations by the method of Brewster, and assuming that they have an energy of 2500–3000 cal (9) higher than the allowed conformations, in first approximation, the molar rotation can be calculated at the different temperatures through the equation

$$[\Phi]_D^T = \frac{-180 + 54\, e^{-E/RT}}{1 + 10\, e^{-E/RT}}$$

where E is 2500–3000 cal and 54 is the average molar rotation for the strained conformations. In this way, between 300 and 400°K, we obtain $\Delta[\Phi]_D/\Delta T = 0.33/0.21$. By applying the same method to the other diastereoisomer (which has 5 conformations with only one strain) we obtain $\Delta[\Phi]_D/\Delta T = -0.14/-0.10$. Both these values have the same sign and are of the same order of magnitude as the experimental values (see Table III).

III. CONFORMATIONAL ANALYSIS IN HIGH-MOLECULAR-WEIGHT ISOTACTIC POLY-α-OLEFINS

As the optical activity per monomeric unit in high-molecular-weight poly-α-olefins is independent from molecular weight, a first attempt was made to calculate the optical rotation per monomeric unit considering each monomeric unit as an independent entity.*

Contrary to what has been observed in the low-molecular-weight paraffins, the experimental values found in this way per monomeric unit of optically

* For a polymer of the type $\;$ ~~~—CH$_2$—^{a}CH$-^b$CH$_2$—^{a}CH$-^b$CH$_2$—^{a}CH$-^b$~~~
   $\qquad\qquad\qquad\qquad\qquad\quad$ |c $\qquad$ |c $\qquad$ |c
   $\qquad\qquad\qquad\qquad\qquad\quad$ R $\qquad$ R $\qquad$ R

the optical activity was calculated with the Brewster method considering for each monomeric unit, beside the nonterminal bonds contained in the R group, the bonds a, b and c.

Table IV Comparison between Experimental and Calculated Molar Optical Rotation for Isotactic Poly-α-olefins

Polymer	$[\Phi]_D^{25}$ experimental[a]	Number of allowed conformations for the monomeric unit	$[\Phi]_D$ calculated		
			[b]	[c]	[d]
I Poly-(S)-3-methyl-1-pentene	+161	3	+40	+180	+240
III Poly-(S)-4-methyl-1-hexene	+288	3	+60	+240	+300
IV Poly-(S)-5-methyl-1-heptene	+68	9	+26	+228	+360

[a] Per mole of monomeric unit (see references 1 and 6)

[b] According to Brewster (9), i.e., averaging among all the allowed conformations.

[c] Average among the rotations of the conformations allowed to the monomeric unit, inserted in left-handed helical sections of the main chain.

[d] Molar rotation, calculated according to Brewster (9), of the allowed conformation having the highest positive molar rotation.

active isotactic poly-α-olefins (having the asymmetric carbon atom in α or β position with respect to the main chain) are in all cases much higher than those which were calculated (Table IV); furthermore, the temperature coefficient of the optical rotation is much higher than that found in the above-mentioned low-molecular-weight paraffins (Table V).

The above discrepancy between experimental and calculated values of the optical rotation is of opposite sense and much larger than that found in the case of the low-molecular-weight paraffins having optical activity of the same order of magnitude. It was taken as an indication that the conformation of each monomeric unit cannot be considered as independent from that of the neighboring ones. An inspection of the allowed conformations for poly(S)-4-methyl-1-hexene (III) shows that, when joining together in the isotactic way monomeric units having positive optical rotation (Figs. 1a and b)* necessarily, the only "allowed" conformation for the principal chain is a left-handed helical conformation.

On the contrary, joining together in the isotactic way the monomeric units having the allowed conformation with negative rotation (Fig. 1c), the only allowed conformation for the main chain is a right-handed helical conformation.

All the conformations of the isotactic main chain including monomeric units having conformations with positive and negative optical activity (i.e., main chains with conformational reversals) have necessarily higher conformational energy (13).

* These two conformations are different only because of the relative positions of the methyl groups that are in γ and δ positions with respect to the principal chain. See the conformational analysis in reference 13.

Table V Comparison of $[\Phi]_D^{25}$ and $\Delta[\Phi]_D/\Delta T$ between Optically Active Poly-α-olefins and Low-Molecular-Weight Model Compounds

Polymer	$[\Phi]_D^{25a}$			Model compound	$[\Phi]_D^{25}$		
	Maximum observed value	Calculated according to Brewster (9)[b]	$\dfrac{\Delta[\Phi]_D^c}{\Delta T}$		Maximum observed value	Calculated according to Brewster (9)[b]	$\dfrac{\Delta[\Phi]_D^c}{\Delta T}$
I Poly(S)-3-methyl-1-pentene	+161	+40	-0.36	XIa (3S,5S)-2,2,3,5-Tetramethylheptane	-97.5	-180	+0.152
III Poly(S)-4-methyl-1-hexene	+288	+60	-0.68	XIb (3R,5S)-2,2,3,5-Tetramethylheptane	+137.8	+120	-0.128
IV Poly(S)-5-methyl-1-heptene	+68	+26	-0.34	XII (S)-5-Ethyl-2,2,3-trimethylheptane	-132	-240	+0.155
				X (S)-3-Ethyl-5-methylheptane	+33.8	+60	-0.065
				XIII (R)-5-Ethyl-2,3-dimethylheptane	+60.2	+105	-0.056

[a] Per mole of monomeric unit.
[b] Average among all the allowed conformations (9).
[c] Between 25 and 110°C.

As previously mentioned, for each (S) monomeric unit inserted in a left-handed helix, two conformations are allowed, while for the same monomeric unit inserted in a right-handed helix, only one conformation is allowed. If the macromolecules have equal terminal groups of suitable structure, considering only the conformations with the lowest conformational energy, only one conformation exists in which the main chain is spiraled in the right sense, while 2^n conformations exist in which the main chain is spiraled in the left-handed sense, n being the number of the monomeric units.

For this reason, an energy difference of about 400 cal/mole of monomeric unit included, respectively, in the more favored or in the less favored helical conformation of the main-chain macromolecule has been roughly estimated (*14–16*). This is one of the reasons for the failure, in this case, of the Brewster calculations based on the equal statistical weight of all the allowed conformations of a single monomeric unit.

These considerations are, however, strictly valid only if we admit that the macromolecules assume rigid helical conformations in which no conformational

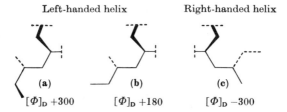

Figure 1   Allowed conformations for the monomeric unit of poly-(S)-4-methyl-1-hexene.

reversals can exist; this is clearly impossible, taking into account that the conformational equilibrium is rapidly achieved at room temperature. Actually, in the above branched poly-α-olefins, despite the fact that each conformational reversal causes an increase in the conformational energy, a number of conformational reversals must be present in each macromolecule in order to minimize the free energy of the polymer (*13, 17*).

The average length of the main chain sections spiraled in the more favored screw sense included between two conformational reversals can be calculated with known methods (*14, 16*) and, independently, starting with the temperature coefficients of the optical rotation (*13*). In the case of poly-(S)-4-methyl-1-hexene, the above-average length included between two conformational reversals, is about 22–28 monomeric units at 300°K and 11–15 at 400°K, the main chain being spiraled correspondingly 91–92% and, respectively, 81–83%, in the thermodynamically more favored screw sense.

According to the above considerations, a much better agreement is found between calculated and experimental values of the optical rotation, admitting that for the polymers I and III, when the asymmetric carbon atoms of the lateral chains have (S) absolute configuration, monomeric units having one or the other of the two allowed conformations with positive optical activity largely prevail (Table IV).

The above considerations also explain the existence, in the polymers, of larger temperature coefficients of the rotation (Table V) which arise not only, as in low-molecular-weight paraffins, from a larger participation of forbidden conformations to the conformational equilibrium by increasing the temperature, but mainly from the change with temperature of the ratio between the number of monomeric units included in sections spiraled in the more favored helical sense and the number of monomeric units included in sections spiraled in the less favored helical sense (13). This fact appears particularly clear when comparing the temperature coefficient of III with that of X. In fact, in (S)-3-ethyl-5-methylheptane (in which, as in III, two positive conformations exist having, respectively, rotatory power of +300 and +180 and one negative having rotatory power of −300) the temperature coefficient of the rotation is less than −0.1, while in III it is about −0.7.

IV. FINAL REMARKS

The results obtained in the present work can be summarized as follows:

(1) As foreseen by Brewster (9) and according to our interpretation of the optical rotation in optically active poly-α-olefins, it has been experimentally proved that conformations exist in branched heptanes having optical activity between 100 and 200.

(2) For reasons which could not be completely clarified, the empirical calculations of optical rotation according to Brewster yield the correct sign, but in general much lower rotation values than that experimentally found when only few allowed conformations having high rotation are predicted. It seems, therefore, that the rotation values obtained by the Brewster method when the ratio between the allowed and forbidden conformation is low, must be taken as largely approximated, errors as large as 100% with respect to the experimental values being possible.

(3) The fact that the experimental values of rotation in some isotactic poly-α-olefins are up to 5 times higher than the values calculated according to Brewster, considering each monomeric unit as a separate entity, is satisfactorily explained, admitting that the rotation per monomeric unit cannot be calculated independently, the allowed conformations of the macromolecules

(which are necessarily prevailingly helical conformations) having in this case remarkably different energy (*13*).

(4) On the same basis, the large differences found in the temperature coefficient of the optical rotation in some isotactic poly-α-olefins and in low-molecular-weight models, having rotation of the same order of magnitude, can be explained.

(5) The soundness of the present views on the conformational analysis both of low- and high-molecular-weight hydrocarbons is confirmed.

ACKNOWLEDGMENT

Work was partially sponsored by the Institute of Macromolecular Chemistry of C.N.R., Milan, Italy.

REFERENCES

1. P. Pino, F. Ciardelli, G.P. Lorenzi and G. Montagnoli, *Makromol. Chem.* **61**, 207 (1963).
2. K. Freudenberg and W. Lwowski, *Liebigs Ann. Chem.* **587**, 224 (1954).
3. F. Ciardelli and P. Pino, *Ricerca Sci.* **34** (II-A), 694 (1964).
4. P.A. Levene and R.E. Marker, *J. Biol. Chem.* **92**, 456 (1931).
5. L. Lardicci, F. Navari, and R. Rossi, *Tetrahedron* **22**, 1991 (1966).
6. P. Pino, *Adv. Polym. Sci.* **4**, 393 (1965).
7. P. Pino and P.L. Luisi, *J. Chim. Phys.* **65**, 130 (1968).
8. P. Pino, P. Salvadori, E. Chiellini and P.L. Luisi, *J. Pure Appl. Chem.* **16**, 469 (1968).
9. J.H. Brewster, *J. Am. Chem. Soc.* **81**, 5475 (1959).
9a. L. Lardicci, R. Rossi, S. Pucci, M. Aglietto, C. Bottighi and P. Pino, *Chim. Ind. (Milan)* **50**, 227 (1968).
9b. P.A. Levene and R.E. Marker, *J. Biol. Chem.* **95**, 3 (1932).
9c. L. Lardicci, P. Salvadori and P. Pino, *Ann. Chim.* **52**, 562 (1962).
9d. P. Pino, S. Pucci, E. Benedetti and P. Bucci, *J. Am. Chem. Soc.* **87**, 3263 (1965).
9e. P. Pino, S. Pucci, E. Benedetti and P. Bucci, unpublished results.
10. S. Pucci, M. Aglietto, P.L. Luisi and P. Pino, *J. Am. Chem. Soc.* **89**, 2787 (1967).
11. A. Abe and M. Goodman, *J. Polym. Sci.* **A1**, 2193 (1963).
12. E. Benedetti, unpublished results.
13. P.L. Luisi and P. Pino, *J. Phys. Chem.* **72**, 2400 (1968).
14. T.M. Birshtein and P.L. Luisi, *Vysokomol. Soed.* **6**, 1238 (1964).
15. T.M. Birshtein and O.B. Ptitsyn, "Conformations of Macromolecules," Wiley (Interscience), New York, 1966.
16. G. Allegra, P. Corradini and P. Ganis, *Makromol. Chem.* **90**, 60 (1966).
17. G. Allegra, P. Ganis and P. Corradini, *Makromol. Chem.* **61**, 225 (1963).
18. S. Pucci, M. Aglietto and P.L. Luisi, *Gass. Chim. Ital.* **100**, 159 (1920).

Quantitative Conformational Analysis of Cyclohexane Systems

J. Reisse

Faculté des Sciences
Université Libre de Bruxelles
Brussels, Belgium

Many works concern the determination of the so-called "A" value for mono-substituted cyclohexane systems (*1, 2*)

$$C_6H_{11}X(a) \;\rightleftharpoons\; C_6H_{11}X(e) \qquad (1)$$

In the case of equilibrium (1), the conformational equilibrium constant, K, is given by the ratio of concentration of each conformation and K is related to $\Delta G°$, $\Delta H°$ and $\Delta S°$ by the familiar expressions

$$K = C_e/C_a \qquad (2)$$

$$-RT \ln K = \Delta G° = -A \qquad (3)$$

$$\Delta G° = \Delta H° - T \,\Delta S° \qquad (4)$$

Since K is defined in terms of concentrations in solution and not in terms of activities, $\Delta G°$ is solvent-dependent.

The values of $\Delta G°$, $\Delta H°$ and $\Delta S°$ are very simply related to $\Delta H_0°$, Q_e and Q_a, where $\Delta H_0°$ (equal to $\Delta E_0°$) is the standard enthalpy difference at $0°K$ and Q_e, Q_a are the partition functions for the two species.

In an ideal gaseous phase, Q is easily factorized in a product of translational, rotational (rigid and internal rotation), vibrational, electronic and nuclear contributions. For equilibrium (1), the translational, electronic and nuclear contributions are obviously the same for the two conformations and may be neglected. We have thus the following expressions:

$$\Delta H° = \Delta E° \text{ (isomerization equilibrium (\textit{3, 4}))} \qquad (5)$$

$$\Delta G° = \Delta E_0° - RT \ln (Q_e/Q_a) \qquad (6)$$

$$\Delta H^\circ = \Delta E_0^\circ + RT^2 \left[\frac{\partial \ln (Q_e/Q_a)}{\partial T} \right] \tag{7}$$

$$\Delta S^\circ = R \ln (Q_e/Q_a) + RT \left[\frac{\partial \ln (Q_e/Q_a)}{\partial T} \right] \tag{8}$$

$$Q_{\text{vib}} = (I - e^{-k\nu/kT})^{-1} \text{ (one vibrational degree of freedom)} \tag{9}$$

$$Q_{\text{rot}} = (8\pi kT/h^2)^{3/2} \pi^{1/2} (I_A I_B I_C)^{1/2} \sigma^{-1} \text{ (rigid rotation—}$$
$$\text{nonlinear molecules)} \tag{10}$$

where $I_A I_B I_C$ is the product of the principal moments of inertia of the molecule and σ is the symmetry number.

In the particular cases of chloro- and bromocyclohexane and corresponding 4,4-dimethyl derivatives, $Q_{\text{(internal rotation)}}$ may also be considered as the same for the two conformations. Nevertheless, this is not always the case, even for two conformations of a monosubstituted cyclohexane.* For example, if the barrier which hinders the internal rotation of the methyl group of methyl-cyclohexane is higher in axial (because of the interaction with 3 and 5 axial hydrogen atoms) than in equatorial position, then $Q_{\text{e(int rot)}}$ must be more important than $Q_{\text{a(int rot)}}$ and gives a positive contribution to the entropy difference. Beckett, Pitzer and Spitzer (5) have considered a higher methyl rotational barrier for the 1,1- and 1,2-dimethylcyclohexane than for the other dimethylcyclohexanes. Only by making this assumption, was it possible to calculate good entropy differences between the various dimethylcyclohexanes.

The expressions (9) and (10) and also the hypothesis concerning the identity (for the two conformations) of translational contributions to Q are valid only in the ideal gas phase (6, 7). The liquid-phase situation is much more complex. Especially, two kinds of degrees of freedom are modified: the translational and the rotational. On the other hand, vibrational degrees of freedom of many monosubstituted cyclohexane systems are practically unaffected by the transition from gas phase to liquid phase or solution in inert solvent. The monohalogenocyclohexanes concerned in this paper are of that kind.

Monohalogenocyclohexanes have been extensively studied from the conformational point of view by various techniques (1, 2, 8, 9, 10). It seems possible to conclude that ΔG° is practically solvent-independent; this observation does not prove that ΔH° and ΔS° are also solvent-independent. An opposite variation of these two terms may exist as in the case of cyclohexanol (11), but it is interesting to note that for bromocyclohexane the ΔH° obtained by the infrared method in gas phase is very near the ΔH° value obtained by nuclear magnetic resonance (nmr) on a solution in carbon disulfide (10). We shall thus

* If the rotation of a substituent is associated with an internal rotational degree of freedom, this substituent is not conical. For instance, CH_3, OH, NO_2, NH_2 are not conical substituents, while H, Cl, Br are.

make the *rough assumption* that we are allowed to discuss our results, qualitatively or semiquantitatively, by means of the classical expression for gas-phase molecules.

I. EXPERIMENTAL RESULTS

We have determined K values at different temperatures for the following systems:

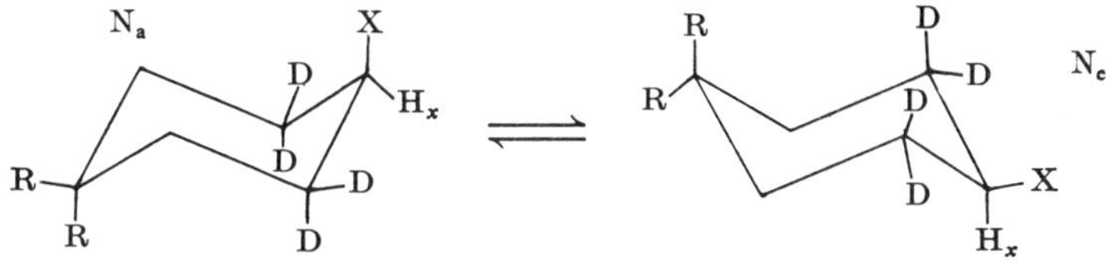

Scheme I

where R = H, CH$_3$ and X = Cl, Br. The method used is nmr in slow-exchange range (−110 to −80°C) and fast-exchange range (−40 to +30°C).

In the slow-exchange range, K is simply obtained by direct integration. The H$_x$ proton, deshielded by the geminated halogen, gives a well-resolved peak for each conformation. The α deuteration avoids all risks of overlap for the two peaks and it is not difficult to remove the errors produced by the differences between saturation factors of the two signals (*12*).

In the fast-exchange range, K is obtained by use of the well-known expression,

$$K = \frac{\delta_e - \delta}{\delta - \delta_a} \tag{11}$$

where δ is the chemical shift of the H$_x$ average signal, and δ_e and δ_a are the chemical shifts of the H$_x$ signal in equatorial and axial conformations, respectively. These two values must be measured at the same temperature as δ. In this work, a linear extrapolation procedure (least-square treatment) has been used from the slow-exchange region, where δ_e and δ_a are directly obtainable, to the fast-exchange region, where δ is measured. In this case too, the use of deuterated compounds is necessary to suppress errors in chemical shift measurements. The experimental technique and data treatment will be extensively described in another paper.

Having K values at different temperatures, it is possible to obtain $\Delta H°$ and $\Delta S°$ using a van't Hoff plot. These results are given in Table I. In each case, chlorocyclohexane excepted, the errors given are statistical errors. In the case of chlorocyclohexane, we have given two extreme values for $\Delta H°$ and $\Delta S°$ obtained by a graphical method using two selected sets of experimental points. In this way, we have an estimation of the superior limit of the possible error. Some values formerly obtained in our research group by the modified Eliel's method are also mentioned, for comparison.

The discrepancy between the results obtained by the two methods is significant and this is in agreement with the observation of Jensen, Beck and Bushweller (2, 12) concerning the A values of numerous cyclohexane compounds. This discrepancy, certainly due to a partial lack of accuracy of the modified Eliel's method, is not sufficient in our opinion to contest the interest of this method. The high precision (reproducibility) and the convenience of

Table I   Conformational Values for Halogenocyclohexanes

	X = Br		X = Cl	
	This work (solvent, CS$_2$)	Eliel method (solvent, decaline)	This work (solvent, CS$_2$)	Eliel method (solvent, decaline)
R = H				
$\Delta G^\circ_{0^\circ C}$[a]	−500	−375	−600 to −535	−410
ΔH°[a]	−450 ± 25	−250 (8)	−430 to −520	−340 (8)
ΔS°[b]	+0.18 ± 0.14	+0.45 (8)	+0.62 to +0.06	+0.25 (8)
R = CH$_3$				
$\Delta G^\circ_{0^\circ C}$[a]	−385		−395	−417
ΔH°[a]	−400 ± 15		−370 ± 20	−240 (16)
ΔS°[b]	−0.05 ± 0.06		+0.09 ± 0.1	+0.65 (16)

[a] ΔG° and ΔH° are given in cal/mole.
[b] ΔS° is given in cal/mole-degree.

the Eliel's method permit solvent studies (with active solvent) on conformational equilibrium. Such studies appear almost impossible from a practical point of view by the "direct nmr method"; the solubility at low temperature, especially in solvents which produce strong solvent interaction appears as a severe limitation. In many cases it is as important to measure the variation of the conformational equilibrium from one solvent to another, as to obtain a very accurate K value in one solvent only. By using synthetic results, we have proved that in many cases Eliel's modified method allows a correct estimation of the sense and the importance of a solvent effect on ΔH° or ΔS°, even if the absolute value of ΔH° or ΔS° in the reference solvent is inaccurate (13).

II. INTERPRETATION OF THE RESULTS

The interest of quantitative conformational studies really appears when an interpretation can be given of the experimental enthalpy or entropy values.

A. Entropy Difference between Axial and Equatorial Conformations

Because in many cases only A values are known, it is usual to consider that $\Delta S°$ is negligible for cyclohexane systems with conical substituents like halogens (14). For many years, we have claimed that the entropy term may be nonnegligible, even for the conical halogen substituent or for the trigonal CH_3 substituent (8, 15, 16). Jensen and co-workers apparently share this opinion (2, 17).

In the case of gas-phase molecules, the rotational entropy (rigid rotation) for a cyclohexane molecule with an equatorial substituent must be higher than for an axially substituted cyclohexane. The product of principal inertia moments of a monosubstituted cyclohexane is effectively always higher for an equatorial conformation than for an axial one (18, 19). The rigid rotation entropy contribution to $\Delta S°$ is temperature-independent. For monohalogeno-

Table II   Vibrational Entropy

	Axial conformation		Equatorial conformation	
Temperature (°K)	ν (cm^{-1})	$S°_{vib}$ (eu)	ν (cm^{-1})	$S°_{vib}$ (eu)
298	686	0.32	726	0.27
	<200?	>2.81?	339	1.21
198	686	0.08	726	0.06
	<200?	>1.32?	339	0.63

cyclohexanes (10), this positive entropy contribution is small and still reduced in the 4,4-dimethyl derivatives.

The vibrational entropy contribution is not so easy to evaluate because all the normal modes of very low frequency are unattributed (20). Table II gives some frequencies and corresponding vibrational entropy contributions for chlorocyclohexane.* The frequencies used are those given by Klaeboe, Lothe and Lunde (20). The entropy calculations were worked out by means of the Einstein's functions tables (7).

The attribution of 725 cm^{-1} and 686 cm^{-1} bands, respectively, to $\nu(C-Cl)_e$ and $\nu(C-Cl)_a$ is unambiguous, but the $\delta(C-C-Cl)$ mode of axial conformation is unknown, perhaps at frequencies lower than 200 cm^{-1} or masked by the strong band at 339 cm^{-1} (20). If we compare equatorial and axial chlorocyclohexane, on one hand, anti and gauche 1-chloropropane on the other, we observe a very good $\nu(C-Cl)$ frequency correlation (equatorial corresponding to anti and axial to gauche). If the same correlation exists for the C–C–X deformation,

* See note added in proof.

the attribution of the $\delta(C-C-X)_a$ used in Table II is unlikely, since $\delta(C-C-Cl)_{gauche}$ for 1-chloropropane is 290 cm^{-1} (21).

Thus, we make the assumption that the contribution to ΔS_{vib}° of the $\delta(C-C-Cl)$ mode is overestimated in Table II. Many other vibrational modes (ring frequencies) must be present below 700 cm^{-1} in the spectrum of chlorocyclohexane (20), with always a possible difference between similar modes in axial and equatorial conformation. These frequencies probably give entropy contributions to ΔS°, but even the sign of these contributions is unknown. If the frequency differences between similar modes are small, each contribution must also be unimportant. Nevertheless, it is impossible to estimate ΔS_{vib}° for chlorocyclohexane. The same remark is valid for bromocyclohexane and for the two 4,4-dimethylhalogenocyclohexanes. It is likely that ΔS_{vib}° is almost the same for halogenocyclohexanes and 4,4-dimethylhalogenocyclohexanes (ν(C–X) is practically unaffected by the 4,4-disubstitution; it is probably the same for the $\delta(C-C-X)$ mode and perhaps the same for ring frequencies).

If we consider our experimental results (Table I), we observe a decrease of ΔS° going from bromocyclohexane to 4,4-dimethylbromocyclohexane and from chlorocyclohexane to 4,4-dimethylchlorocyclohexane. This behavior is in agreement with the present discussion. The positive rotational entropy contribution is reduced by the disubstitution in position 4 and the vibrational entropy contribution is almost constant from monohalogenocyclohexane to 4,4-dimethylhalogenocyclohexane. Moreover, the experimental results seem to prove that the vibrational entropy contribution to ΔS° is very small. The constancy of ΔH° and ΔS° in the temperature range used in the course of this work is also in agreement with this last conclusion. Effectively, only the vibrational partition function contributes to the variation of ΔS° and ΔH° with temperature. In the case of 4,4-dimethylbromocyclohexane for which the precision of the experimental results is particularly high, we have calculated ΔH° and ΔS° making use of the following:

(1) the whole of the temperature range (see results in Table I);

(2) only the integration values obtained in the low-temperature range (−110 to −80°C).

The two sets of results are the same within the limits of errors. It is easy to infer from the values noted in Table II, the possible influence of two very distinct low-frequency modes upon the difference between $\Delta S_{298°K}^{\circ}$ and $\Delta S_{198°K}^{\circ}$.

B. Enthalpy Difference between Axial and Equatorial Conformations

The small contribution of the vibrational partition function to ΔS° and ΔH° is a necessary condition to make use of the van't Hoff treatment of data. Nevertheless, it is not a sufficient condition to consider that $\Delta H^{\circ} = \Delta E_0^{\circ}$; ΔE_0° can be calculated by plotting $\ln K(\pi_e/\pi_a)$ versus $1/T$ (22), where π is given by expression (12).

$$\pi = \prod_i [1 - \exp(-h\nu_i/kT)] \tag{12}$$

To be able to calculate ΔE_0°, it is once more necessary to know all the normal modes, and essentially those of low frequency. It is possible to estimate a maximum order of magnitude of this correction, using the frequencies given in Table II and remembering that the difference between $\delta(C-C-X)_e$ and $\delta(C-C-X)_a$ is certainly overestimated. In the case of chlorocyclohexane we have obtained two limit values of ΔH°, -430 and -520 cal/mole, respectively. The corresponding values of ΔE° are -345 and -425 cal/mole. The probably overestimated correction has a value of approximately 100 cal/mole. (See note added in proof.)

In our particular case, for reasons we previously discussed, comparing chlorocyclohexane and 4,4-dimethylchlorocyclohexane or corresponding bromo compounds, we may probably say that the lower absolute value of ΔH° for a 4,4-dimethyl compound would also be present on a ΔE_0° scale. Is it a sufficient reason to try to give an explanation in terms of energy calculation as it appears in the recent work by Allinger et al. (23)? It should be noted that all the conformational energy calculations on complex molecules such as cyclohexane, by semi-empirical methods (23 and references cited) or ab initio method (24) are made for completely quiescent molecules. Allinger et al. (25) give good reasons to calculate energetic values at 298°K and not at 0°K, but they say, "Thus, it is necessary to assume that the other contributions to zero point energy will be proportional to the number of bonds present for each type and proportional for each branch in the chain and hence will cancel out in the present treatment." Thus, in such calculation, the zero-point vibrational energy (ZPVE) is considered as the same for the two conformations.

C. Zero-Point Vibrational Energy; Difference between ΔE_0° and $\Delta E_{0(\text{electronic})}^\circ$

We have the general relation $\Delta E_0^\circ = \Delta E_{0(\text{elect})}^\circ + \Delta E_{0(\text{vib})}^\circ$. Zero-point vibrational energy (ZPVE) is an important term. For example, ZPVE has a value of 107 kcal/mole in the case of cyclohexane (26). This term is easily calculated if all the fundamental frequencies are known, since

$$\text{ZPVE} = \tfrac{1}{2} \sum_i h\nu_i \tag{13}$$

For monohalogenocyclohexanes, the sum must be taken over 48 frequencies. This number is 66 in the case of 4,4-dimethylmonohalogenocyclohexanes. Once more, the absence of a complete vibrational study of the molecules, concerned in this work, avoids the estimation of ZPVE. The importance of

ZPVE is well-recognized in the calculation of excess enthalpy of cycloalkanes (*26, 27, 35*), or in the calculation of formation energy of alkanes (*28, 29*).

An interesting influence of the ZPVE term can be found in the work of Allen (*30*). This author has estimated the strain energy (relative to paraffins) associated with a cyclane ring. In the case of cyclopentane, this strain energy has a value of 6.0 kcal/mole at 25°C; 7.3 kcal/mole at 0°K and 10.5 kcal/mole for the motionless molecule. Thus, the corrections for thermal energy and ZPVE yield a larger strain energy.

Generally, the possible difference between ZPVE of various conformations is neglected. It is well known that for conformations, the sum rule ($\sum_i \nu_i^2$) of Mizushima *et al.* (*31*) works very well. Bernstein and Pullin (*32*) have proposed an empirical sum rule ($\sum_i \nu_i$) for homologs and this empirical sum rule is also quite well satisfied for conformations. For instance, in the case of 1,2-difluoro-ethane (*33*), the term $\sum_i \nu_i$ is equal to 24,920 cm^{-1} for the gauche isomer and 24,739 cm^{-1} for the anti. If the empirical sum rule was perfectly satisfied for conformations, ZPVE of any conformation of one molecule would be evidently the same. Unfortunately, as we can see for 1,2-difluoroethane, the difference between $(\sum \nu_i)_{\text{gauche}}$ and $(\sum \nu_i)_{\text{anti}}$ is approximately 200 cm^{-1}, and in this case, if the vibrational assignment is correct, the difference in ZPVE between the two conformations is *more than* 250 cal/mole. Very recently, Craig and Overend (*34*) have published a new vibrational study of *cis*- and *trans*-1,2-difluoro-ethylenes and a new estimation of the thermodynamic functions of these compounds. For the equilibrium *cis* $\rightleftharpoons$ *trans*, they give the following expression:

$$\Delta E^\circ_{0(\text{elect})} = \Delta H^\circ_{650°K} - \Delta H^\circ_{650°K(\text{vib})} - \Delta E^\circ_{0(\text{vib})}$$

$$= 930 \qquad - 144 \qquad + 300 = 1086 \text{ cal/mole}$$

As we can see, the unequal population of vibrational levels gives a correction of -144 cal/mole and the difference of the ZPVE for the two stereoisomers which gives the $\Delta E^\circ_{0(\text{vib})}$ term is 300 cal/mole. On this basis, it seems unaccept-able to postulate $\Delta E^\circ_{0(\text{vib})} = 0$ for conformational equilibria such as those studied in the course of this work. Moreover, there are no serious reasons to postulate $\Delta E^\circ_{0(\text{vib})}$ constant in a series of compounds such as halogenocyclo-hexanes. In these conditions, it is perhaps hazardous to justify sequences of ΔG° or ΔH° in terms of bond lengths and polarizabilities of halogen (*2*).

On the other hand, it seems also dangerous to interpret, on a purely electronic energy basis, the lowering of ΔH° by passing from monohalogeno-cyclohexanes to 4,4-dimethyl derivatives. Even if the low-frequency regions of the two kinds of compounds seem similar, it is unsafe to consider as impossible a situation like the following:

$$\left(\sum_{\nu=1}^{\nu=48} \nu_i\right)_{\text{axial}} = \left(\sum_{\nu=1}^{\nu=48} \nu_i\right)_{\text{equatorial}} + A$$

for bromocyclohexane and

$$\left(\sum_{\nu=1}^{\nu=66} \nu_i\right)_{\text{axial}} = \left(\sum_{\nu=1}^{\nu=66} \nu_i\right)_{\text{equatorial}} + B$$

for 4,4-dimethylbromocyclohexane, with a difference between $|A|$ and $|B|$ of 40 cm^{-1}. Such a difference is sufficient to justify the observed effect on a ZPVE basis.

III. CONCLUSIONS

At this time, experimental measurements of conformational equilibrium can be made with great accuracy and precision. A difference of 50 or 100 cal/mole between two $\Delta H°$ values is probably significant. On the other hand, the precision of energy calculations increases and it becomes tempting to justify very small values of energy or free-energy differences in terms of nonbonded interactions, bond lengths, valence angles or torsional angles deviations.

A good example of such an attempt may be found in reference 23. In this work, the free-enthalpy differences between the two conformations of 1-chloro-1-methylcyclohexane is calculated. The obtained value (1.03 kcal/mole) is different from the value estimated if additivity of free enthalpy of the methyl group and the chlorine atom is assumed (1.21 kcal/mole). For Allinger et al. (23), "The difference is not attributable to round-off error in the calculations but is a real calculational difference. It is in excellent agreement with the experimental value but the experimental measurements were not sufficiently precise to definitely exclude the additivity value."

After the preceding discussion, we consider that without a preliminary vibrational study and a calculation of the vibrational contributions to enthalpy differences, conformational effect lower than 200 cal/mole (and perhaps more) may not be safely interpreted, and there is, for us, a present limitation of conformational analysis.

Note added in proof. Recently, P. Klaeboe (*Acta Chem. Scand.* **23**, 2641 (1969)) has published a new vibrational assignment for halogenocyclohexanes. Using the new frequency values for chloro- and bromocyclohexanes, the ZPVE of axial conformation appears *higher* than for the equatorial one; the difference between the ZPVE terms being approximately −100 cal/mole for the two derivatives. On the other hand, $(\Delta S°_{\text{vib}})_{298°K}$ seems to be positive and similar in magnitude for the two compounds (0.30 eu for chlorocyclohexane and 0.24 eu for bromocyclohexane). The thermal contribution to $\Delta H°_{298°K}$ is negligible, and in the covered temperature range, $\Delta H°$ and $\Delta S°$ can be considered as constant.

ACKNOWLEDGMENTS

This work was performed in collaboration with Professor Oth (Eidgenössische Technische Hochschule), Dr. Gilles (Union Carbide ERA), and Dr. Stien and Mr. Leit (Brussels University).

REFERENCES

1. J.A. Hirsch, *in* "Topics in Stereochemistry," N.L. Allinger and E.L. Eliel, eds., p. 199, Wiley (Interscience), New York, 1967.
2. F.R. Jensen, C.H. Bushweller and B.H. Beck, *J. Am. Chem. Soc.* **91**, 344 (1969).
3. K.B. Wiberg, "Physical Organic Chemistry," Wiley, New York, 1963.
4. H.F. Bartolo and F.D. Rossini, *J. Phys. Chem.* **64**, 1685 (1960).
5. C. W. Beckett, K.S. Pitzer and R. Spitzer, *J. Am. Chem. Soc.* **69**, 2488 (1947).
6. J.E. Leffler and E. Grunwald, "Rates and Equilibria of Organic Reactions," Wiley, New York, 1963.
7. G.J. Janz, "Thermodynamic Properties of Organic Compounds," Academic Press, New York, 1967.
8. J. Reisse, J.C. Celotti and G. Chiurdoglu, *Tetrahedron Letters* **1965**, 397.
9. E.L. Eliel and R.J.L. Martin, *J. Am. Chem. Soc.* **90**, 689 (1968).
10. J. Reisse, M.L. Stien, J.-M. Gilles and J.F.M. Oth, *Tetrahedron Letters* **1969**, 1917.
11. J. Reisse, J.C. Celotti, R. Ottinger and G. Chiurodoglu, *Chem. Comm.* **1968**, 752.
12. F.R. Jensen and B.H. Beck, *J. Am. Chem. Soc.* **90**, 3252 (1968).
13. G. Ransbotyn, J.C. Celotti, R. Ottinger and J. Reisse, to be published.
14. E.L. Eliel, N.L. Allinger, S.J. Angyal and G.A. Morrison, "Conformational Analysis," Wiley, New York, 1965.
15. J.C. Celotti, J. Reisse and G. Chiurdoglu, *Tetrahedron* **22**, 2249 (1966).
16. G. Ransbotyn, R. Ottinger, J. Reisse and G. Chiurdoglu, *Tetrahedron Letters* **1968**, 2535.
17. B. Rickborn and F.R. Jensen, *J. Org. Chem.* **27**, 460 (1962).
18. M. Larnaudie, Thesis, Paris, 1953.
19. L. Pierce and J.F. Beecker, *J. Am. Chem. Soc.* **88**, 5406 (1966).
20. P. Klaeboe, J.J. Lothe and K. Lunde, *Acta Chem. Scand.* **10**, 1465 (1956).
21. N.T. McDevitt, A.L. Rozek, F.F. Bentley and A.D. Davidson, *J. Chem. Phys.* **42**, 1173 (1965).
22. H.J. Bernstein, *J. Chem. Phys.* **17**, 258 (1949).
23. N.L. Allinger, J.A. Hirsch, M.A. Miller and I.J. Tyminski, *J. Am. Chem. Soc.* **91**, 337 (1969).
24. J.R. Hoyland, *J. Chem. Phys.* **50**, 2775 (1969).
25. N.L. Allinger, J.A. Hirsch, M.A. Miller, I.J. Tyminski and F.A. Van Catledge, *J. Am. Chem. Soc.* **90**, 1199 (1968).
26. M. Bixon and S. Lifson, *Tetrahedron* **23**, 769 (1967).
27. K.B. Wiberg, *J. Am. Chem. Soc.* **87**, 1070 (1965).
28. T.L. Cottrell, *J. Chem. Soc.* **1948**, 1448.
29. K.S. Pitzer and E. Catalano, *J. Am. Chem. Soc.* **78**, 4844 (1956).
30. T.L. Allen, *J. Chem. Phys.* **31**, 1039 (1959).
31. S.I. Mizushima, T. Shinamouchi, I. Nakagawa and A. Miyake, *J. Chem. Phys.* **21**, 215 (1953).
32. H.J. Bernstein and A.D.E. Pullin, *J. Chem. Phys.* **21**, 2188 (1953).
33. P. Klaeboe and J.R. Nielsen, *J. Chem. Phys.* **33**, 1764 (1960).
34. W.C. Craig and J. Overend, *J. Chem. Phys.* **51**, 1127 (1969).
35. S. Lifson and A. Warshel, *J. Chem. Phys.* **41**, 5116 (1968).

New Concepts on a Synthesis of Higher-Membered Catenanes and Knots and Model Investigations

Gottfried Schill
Chemisches Laboratorium der Universität
Freiburg, Germany

My talk apparently has nothing to do with the subject of this convention, since I will not be reporting on investigations of conformational transformations, conformational equilibria or similar problems. In a certain sense, however, the synthesis of catenanes and knots is a conformational problem, in that a particular conformation of the precursors to be cyclized is an absolute prerequisite for the successful synthesis of these compounds. The conformation of molecules composed of long bifunctional chains and macrocycles, which can serve as starting materials for the synthesis of catenanes can, for example, be influenced by alteration of the solvent. Investigations on the statistical synthesis of catenanes have, however, shown, that the equilibrium between intraanular and extraanular conformations of such compounds lies largely on the extraanular side.

The problem of a directed synthesis of catenanes and knots—which ought to give higher yields than a statistical synthesis—consists in establishing a definite conformation by attachment of additional bonds to the molecule being cyclized. In a later stage of the synthesis, these bonds must be ruptured again. In accordance with this principle of synthesis, an intraanular triansa compound I, for example, can be prepared, which may be designated as a precatenane. A [2]-catenane II is obtained, then, by rupture of the bonds between the double bridged system and the aromatic nucleus. Probably only in this or similar manner, is it at all possible to prepare higher-membered catenanes and knots in satisfactory yield, since the probability of their formation is statistically very small.

229

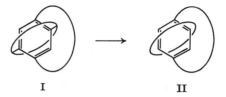

I   II

Scheme I

With the extension of the principle of directed synthesis, a [3]-precatenane V can be prepared in one of two ways:

(1) By means of intraanular attachment of two double-bridged systems onto a cyclophane IV.
(2) By dimerization of an intraanular diansa compound III, which has two chains with terminal functional groups.

A [3]-catenane VI would be expected to result by cleavage of the bonds between the aromatic nuclei and the double-bridged systems.

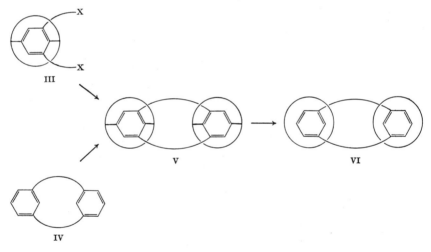

Scheme II

I should like to report on experiments made with both procedures. To carry out the first procedure, we planned the synthesis of a [3]-precatenane by attachment of two double-bridged systems to a tetrahydroxy-*m*-cyclophane VII. In preliminary experiments it was to be established whether two double-bridged systems could be attached to bisdihydroxyphenyl alkanes of type VIII at all.

In these preliminary experiments, the tetrahydroxy compounds VIIIa–c, with n equal to 2, 6 and 12, were ketalized with 1,21-dichloro-heneicosane-11-one (IXa–c), nitrated in the 5 position (Xa–c), reduced to the amines (XIa–c), and then cyclized under high dilution conditions.

A further purpose of the preliminary experiments was to elucidate the question as to whether cyclization of bisamines XIa–c leads only to bisdiansa compounds of type XII or if, perhaps, cross-cyclization also occurs. The latter phenomenon takes place if an alkyl halide residue does not react with the amino group of the same benzene ring, but with that of the juxtaposed benzene nucleus. Since such an effect is only possible if both halves of the molecule are able to come sufficiently close to one another, the length of the polymethylene bridge connecting both benzene rings becomes decisive. To what extent both halves of the molecule can approach one another with a given chain length

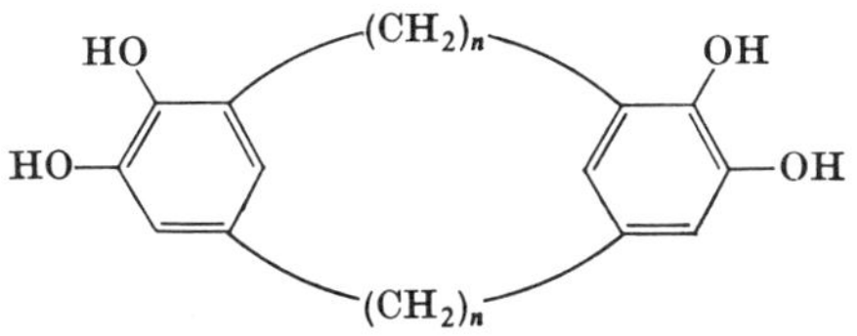

VII

Scheme III

can be estimated from molecular models, but obviously the conformation of the paraffin chain under the conditions of cyclization plays an important role.

The results of the cyclization reactions are given in Table I. With varying yields, a main reaction product was found in all cases. A small quantity of a second reaction product could be isolated in the case of diamine XIc.

A structure elucidation for the cyclization products was achieved on the basis of mass spectra.* In the case of the cyclization product XIIa, a fragmentation between the two methylene groups is particularly favored, because in this case benzyl cations can form. If the product of the cyclization has the structure shown, a fragmentation ion with half of the molecular weight (m/e = 426) is to be expected, and this is actually the case. The peak with half the molecular weight represents the base peak.

That which was said for the cyclization product XIIa is equally valid for the other cyclization products, with the qualification that in these cases fragment ions, which are formed by ruptures at the benzyl position in the polymethylene chain which binds the two halves of the molecule together, are to be expected with less intensity because a fragmentation is less favored

* The mass spectrometric studies were done by Walter Vetter, Hoffmann-La Roche & Co., Bale.

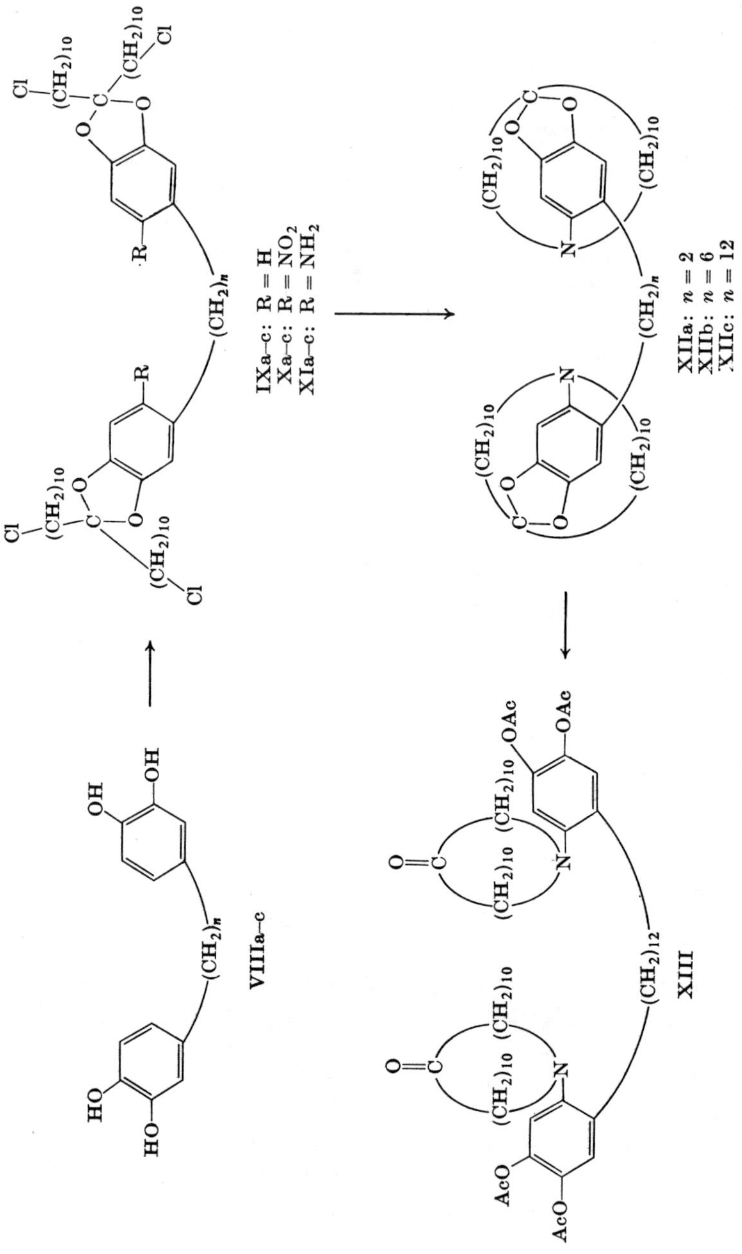

IXa–c: R = H
Xa–c: R = NO_2
XIa–c: R = NH_2

XIIa: $n = 2$
XIIb: $n = 6$
XIIc: $n = 12$

VIIIa–c

XIII

Scheme IV

here. Therefore, by evaluation of the mass spectra, the structures of the products of the cyclization reactions could be confirmed as bisdiansa compounds XIIa–c. The acid-catalyzed hydrolysis of the bisdiansa compound XIIc and subsequent acetylation yielded the tetraacetate XIII.

The byproduct of cyclization of diamine XIc represents a monomer, too. It follows from this that the structures XIV, XV or XVI could be considered for this compound. Compounds XIV and XV represent stereoisomers, which differ from one another in the manner of *cis-trans* isomerism. Ketal hydrolysis and subsequent acetylation yielded a tetraacetate which might have structure XVII, but due to the small amounts, no definite structure assignment could be made. For steric reasons, we are assuming that compound XIV is present.

For the synthesis of a [3]-precatenane using the second procedure, we converted the diansa compound XVIII (*1*) to the bis-*p*-toluenesulfonamide

Table I   Cyclization Results of Some Diamines

Reacted diamine	Cyclization products	Yields (%)[a]
XIa	XIIa	25
XIb	XIIb	7
XIc	XIIc	27
	XIV, XV or XVI	2

[a] Referred to the corresponding dinitro compounds.

XIX. In this compound, the chain and the aromatic nucleus are intraanularly fixed by the ketal grouping and the limiting length of the polymethylene bridges. The cyclization of XIX with the starting dibromide XVIII in a molar ratio 1:1, using the high dilution method, gave a mixture of compounds XX and XXI in 38% yield, uniform in thin-layer chromatography. Detosylation and acetylation gave the diacetyl derivatives XXII and XXIII. When the method established in the synthesis of [2]-catenanes was employed (*2*), the cleavage of the chemical bonds between the aromatic nuclei and the double-bridged systems led to an apparently uniform mixture of [3]-catenanes XXIV and XXV (*3*).

A directed synthesis, as applied heretofore, can be used for building up higher catenanes. However, the individual macrocycles are not then arranged in a continuous chain, but, rather, they are bound topologically by means of a simple macrocycle. The method cannot be applied for synthesizing a chain from macrocycles.

In evaluating these experiments, we have tried to discover the mechanism of syntheses involved in more complicated systems as a knot, a doubly wound

catenane, or a chain of macrocycles. I should like here to limit the discussion
of synthesis possibilities to these relatively simple systems and ignore the
synthesis of more complicated molecules.

According to the principle of fixed conformations, a single diansa compound
XXVI should be suitable as a starting material for the compounds mentioned.
In this molecule, a double-bridged system is bonded through a ketal grouping

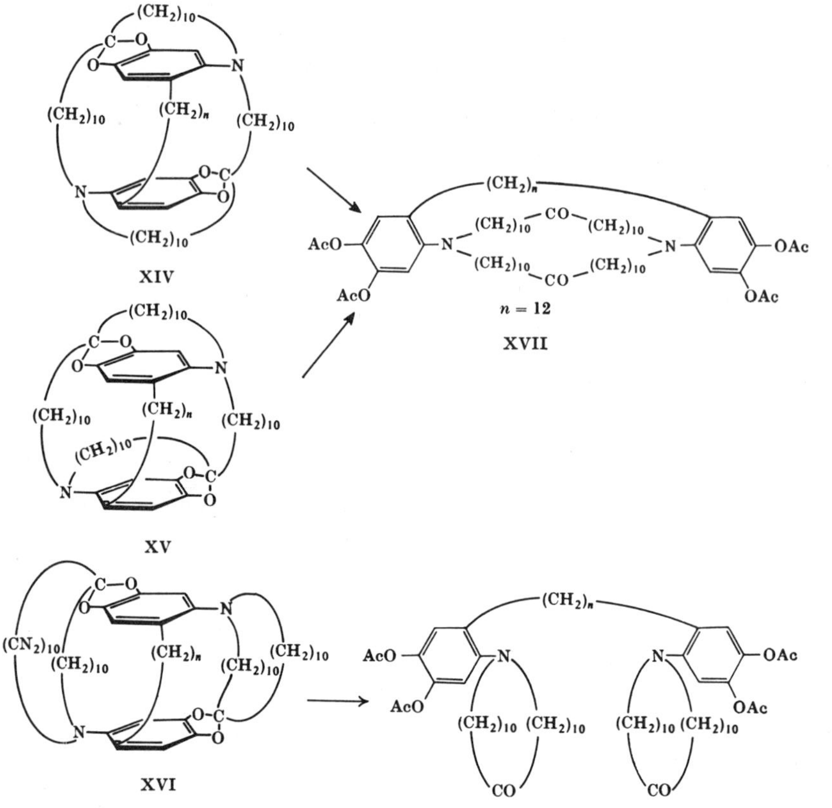

Scheme V

and two amino groups to the aromatic nucleus. Because of the ketal grouping
and the limiting length of both bridges, the latter are located on different
sides of the benzene nucleus and are rigidly fixed. In the 4 and 7 positions,
moreover, two long-chained alkyl fragments with functional end groups are
attached.

The synthesis of a knot should be accomplished by the following reaction
sequence: Alkylation of an amino group by an alkyl halide residue in *meta*
position leads to triansa compound XXVII. Two topologically isomeric

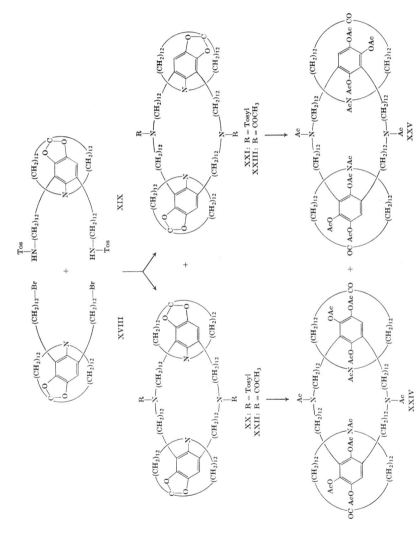

Scheme VI

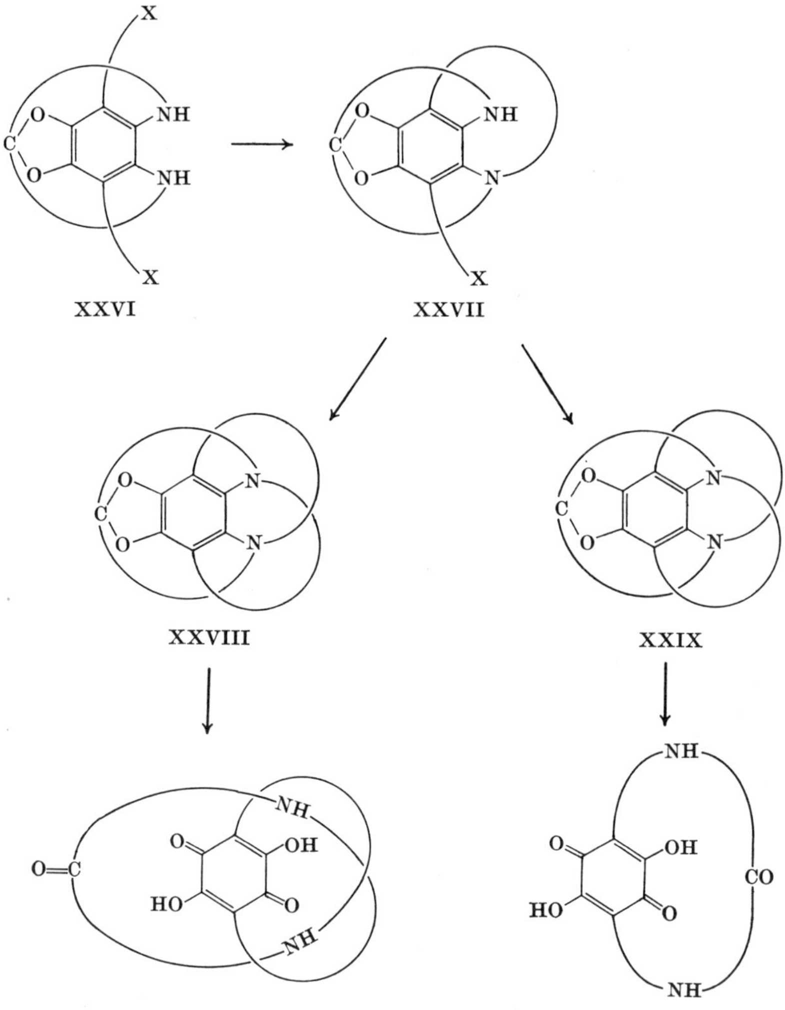

Scheme VII

compounds **XXVIII** and **XXIX** are formed as the other amino group is alkylated by the alkyl halide residue still present. The rupture of the ketal bonds and the aryl–nitrogen bonds yields, in one case, a simple macrocycle **XXXI** and, in the other, a knot **XXX**.

In order to prepare a doubly wound catenane, one could fasten two molecules of diansa compound **XXVI** together, whereby two isomers, **XXXII** and **XXXIII** would result. The closing of both outer bridges leads to the topological isomers **XXXIV** and **XXXV**. One of these (**XXXV**), after rupture of the ketal and the aryl–nitrogen bonds, yields a doubly wound catenane **XXXVIII**; ‧ the other (**XXXIV**), two separate macrocycles **XXXVI** and **XXXVII**. By dimerization of diansa compound **XXVI**, stereoisomer **XXXIII** (one anti-

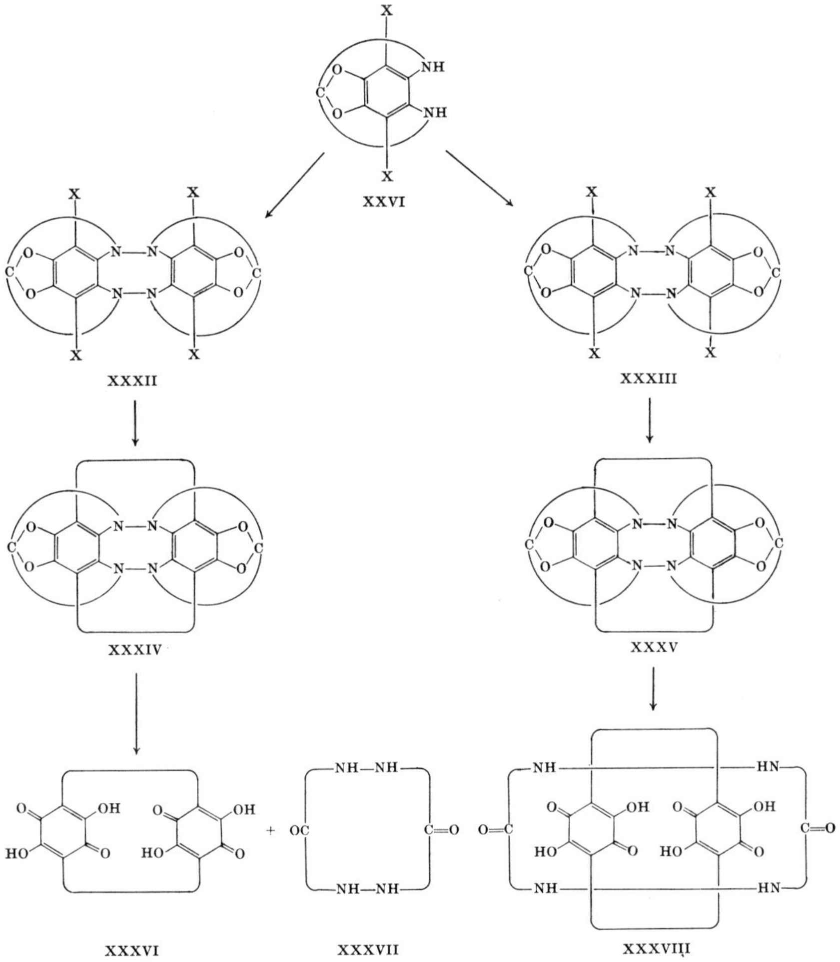

Scheme VIII

pode) can be obtained as a sole product if an optically active diansa compound **XXVI** is used.

The synthesis of a continuous chain of macrocycles, attached as in a chain, can be illustrated using the diansa compound **XXVI** as starting material. The intermolecular cyclization of two molecules results in compound **XXXIX**,

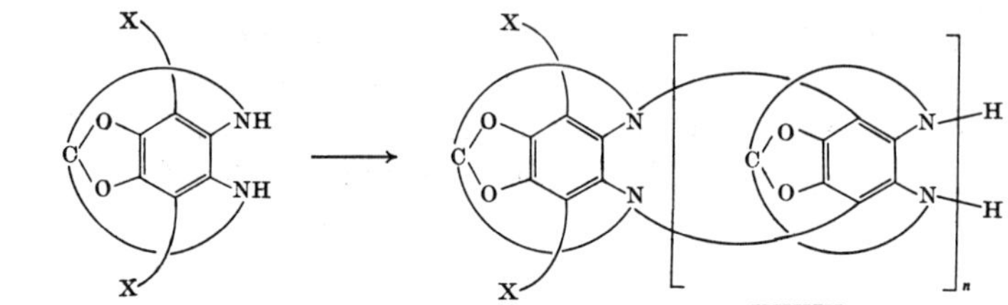

Scheme IX

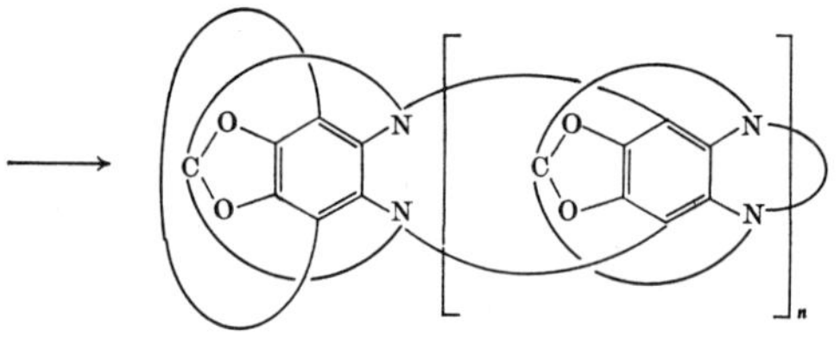

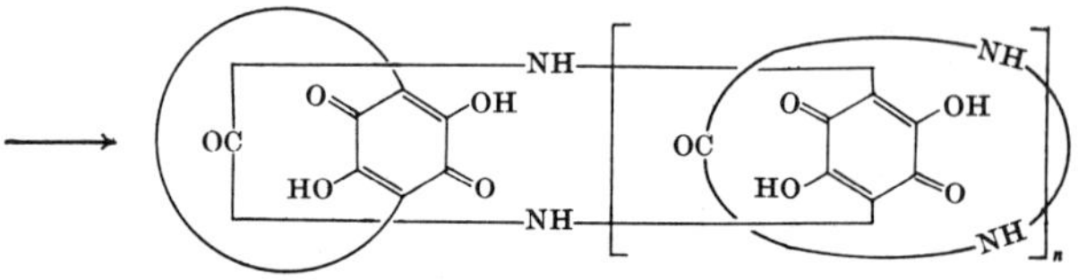

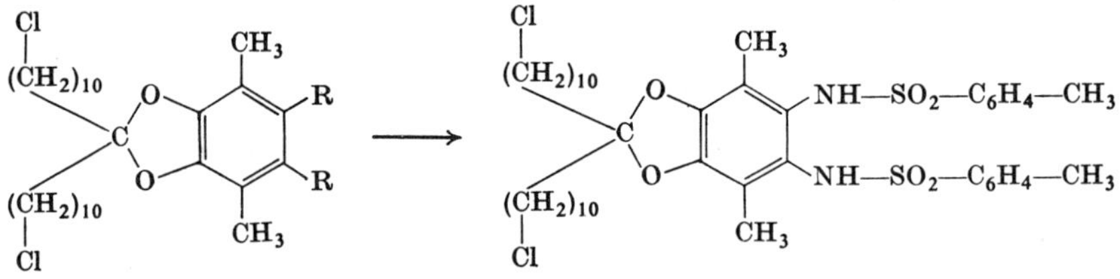

XLII: R = H
XLIII: R = NO₂
XLIV: R = NH₂

XLV

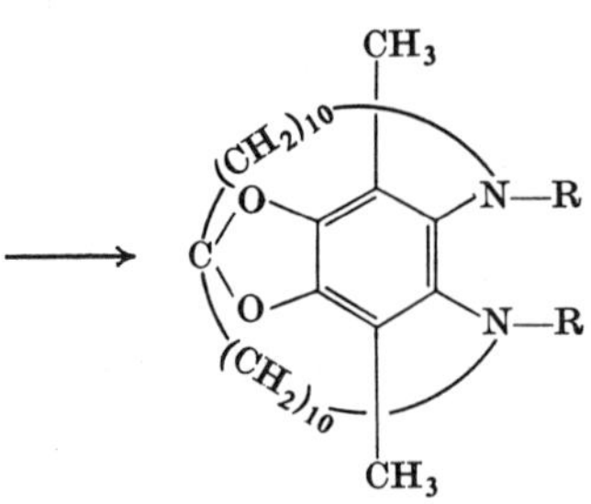

XLVI: R = SO₂—C₆H₄—CH₃
XLVII: R = H

Scheme X

to which further diansa compounds, by repetition of the cyclization step, can be connected. By closing the first and the last ring, one obtains a preoligocatenane XL. The cleavage of the bonds between the aromatic nuclei and the double-bridged systems yields an oligocatenane XLI. In such a molecule, the macrocycles are arranged like the links of a chain.

A diansa compound XXVI of 5,6-diamino-benzodioxole, with two long-chain residues carrying functional end groups, represents the key compound for all syntheses discussed. We have not yet synthesized such a diansa compound. However, in preliminary investigations, a diansa compound XLVII could be prepared, which carries methyl groups in place of long-chain residues.

For this purpose, 3,6-dimethyl-pyrocatechole was ketalized with 1,21-dichloroheneicosan-11-one (XLII) and nitrated to the dinitro compound XLIII. Reduction yielded diamine XLIV, which was converted to the ditosylate XLV. After preparing the corresponding diiodide, the diansa compound XLVI was isolated in about 30% yield by cyclization in DMSO under high dilution conditions. Detosylation yielded the diamine XLVII.

So far, the detachment of the double-bridged system from the aromatic nucleus has not been achieved. However, from our experience to date, we believe that the synthesis—albeit with difficulties—can be carried further, and, thus, the building up of a knot, a doubly wound catenane, and, possibly, too, of a higher unbranched catenane is feasible.

ACKNOWLEDGMENTS

My co-workers, Juan Boeckmann, Raimund Henschel, Klaus Murjahn and Clemens Zürcher contributed to the investigations mentioned. To all of them I would like to express my gratitude.

REFERENCES

1. G. Schill and H. Zollenkopf, *Liebigs Ann. Chem.* **721**, 53 (1969).
2. G. Schill, *Chem. Ber.* **100**, 2021 (1967).
3. G. Schill and C. Zürcher, *Angew. Chem.* **81**, 996 (1969); *Angew. Chem. Intern. Ed. Engl.* **8**, 988 (1969).

Conformational Analysis of Carbonium Ions

Paul von Ragué Schleyer
Department of Chemistry
Princeton University
Princeton, New Jersey

Although carbonium ions are the most extensively investigated organic intermediates, practically no direct experimental structural information is available. x-Ray analyses have only been reported on a few highly stable, aryl-substituted cations (*1, 2*). During the last decade it has been possible by using strong acid media to prepare simple tertiary and secondary carbonium ions as long-lived species (*3,4*). However, x-ray structural data on such species are not available. Recently, techniques have been developed so that the Raman spectra of stable cations can be determined (*5*), but full vibrational analyses on even the simplest species have not yet been accomplished (*6*).

The available evidence on the structure of carbonium ions is indirect (*1, 2*). It is inferred from isoelectronic boron compounds, e.g., $(CH_3)_3B$, that carbonium ions should be planar, but such recourse to analogy can be dangerous. Borane, BH_3, exists as a bridged dimer, but it seems unlikely that the methyl cation CH_3^+ will dimerize. Preferred planarity of carbonium ions is strongly supported by work with bridgehead-substituted polycyclic systems where the cation is constrained to nonplanar arrangements because of ring strain (*2*). The rates of solvolysis of bridgehead derivatives are generally strongly depressed (*2, 7*). A third line of evidence is based on theory; cations with three identical substituents should prefer trigonal, sp^2-hybridized arrangements (*1, 2, 8*). There is no experimental information at all concerning many fundamental structural and conformational questions involving carbonium ions. We have attempted to answer these questions by calculation using two basically different theoretical approaches. The first is based on computer conformational analysis (the Westheimer - Hendrickson - Wiberg - Allinger method) (*9*) and has been applied to the study of the solvolytic reactivity at the bridgehead positions of polycyclic hydrocarbons (*10*). The second approach, based on both semi-empirical and ab initio molecular orbital calculations, has been used to study the smallest carbonium ions (11–13).

I. THE METHYL CATION

Structural problems are encountered already with the methyl cation, a high-energy species (1) which has probably never been observed outside a mass spectrometer. While it would be astonishing if the methyl cation were anything but a planar, trigonal species, it is not immediately obvious what effect distortion would have on the preferred geometry. In fact, the possibility has been suggested that the 7-norbornyl cation (I), because of the rather small (~95°) C–C+–C angle, might have a nonplanar arrangement around the carbonium ion center (14). This postulated nonplanarity might account for the observation that solvolysis of tagged 7-norbornyl derivatives gives predominate retention of configuration, an unusual stereochemical result (14).

I

In order to test this suggestion, we performed a number of calculations on the methyl cation (the simplest reference cation model), the isopropyl cation, and the 7-norbornyl cation (I) itself (11,12). All methods of calculation for all of the species studied gave the same conclusion: carbonium ions, despite distortion, strongly prefer planarity. In fact, the smaller the angle involving the other three bonds at the cation center becomes, the more difficult it is to bend the remaining C+–R bond out of the plane. As electron-deficient species, carbonium ion geometries tend to be dominated by internuclear repulsion terms (15). When a cation is bent out-of-plane, the internuclear distances become shorter, and the interactions worse. This effect is magnified by distortion of one of the angles to small values.

II. CALCULATION METHODS

Before proceeding, it would be best to make some general remarks about the calculation methods employed in this work. These represent various degrees of refinement. These methods start with nuclei and electrons, but make no explicit assumptions concerning bonding. The *semi-empirical* schemes (CNDO, INDO, MINDO, NDDO, etc.) (16) neglect inner-shell electrons and reduce the number of protons in the nuclei appropriately. The interactions of the remaining electrons and cores are considered, but a number of simplifications are introduced; some of the integrals, especially those representing electron repulsions, are neglected. In order to compensate for this neglect, the remaining parameters must be adjusted semi-empirically. The virtue of such methods is their relative computational simplicity; even relatively large molecules can be treated.

In the ab initio methods, as approximations to the Hartree-Fock scheme, all electrons are considered and all integrals are evaluated. There are many variations, some using minimal and some more extended basis sets. In principle, the larger the basis set, the closer the approach to the Hartree-Fock limit. Obviously, when a large basis set is used, the complexity (and the cost) of the calculation is increased enormously, and only the smallest molecules (such as the methyl cation) can be treated, if a reasonably complete geometry search is to be carried out. Since the calculated energies are highly dependent on the structural parameters, it is important that the effect of bond-length variation, etc., be assessed.

A healthily skeptical attitude about the results of such calculations should be maintained. It is instructive to subject the method being used to checks on systems where experimental data is available. Also, comparisons of the various methods serve to provide a calibration of their reliability. Some published results have not been subjected to these tests; reported data seem so improbable as to discredit the whole theoretical approach in the eyes of the general chemical public. We believe that theory, carefully applied, can provide much insight into phenomena not yet amenable to direct experimental investigation, and can often reveal problems unsuspected from currently available data.

III. THE ETHYL CATION

What is the structure of the ethyl cation? Are bridged (protonated ethylene) or classical forms preferred? What is the barrier to rotation in the classical species? The answers to these questions illustrate important limitations in the semi-empirical calculation methods. If local symmetry around the methyl and CH_2^+ groups is assumed, both ab initio and semi-empirical schemes predict a zero rotational barrier in the classical ethyl cation. This is a consequence of symmetry; all sixfold barriers are known to be small ([17]). If the methylene group is deformed to a tetrahedral geometry, a threefold ethanelike barrier results ([18]), the magnitude of which depends somewhat on the method of calculation.

Even qualitative agreement among various molecular orbital schemes is lost when the preferred structure of the ethyl cation is considered. According to semi-empirical results, the bridged ethyl cation should be considerably more stable than the classical arrangement, e.g., in NDDO, by 33.2 kcal/mole ([12]).

Quite the opposite is indicated by ab initio calculations, which favor the classical ethyl cation structure (*12, 13*) by 9–12 kcal/mole! The latter results, based on sounder calculation methods, are to be regarded as the more reliable. The little available experimental evidence, suggesting that an appreciable barrier to the rearrangement of tagged ethyl derivatives exists (*19*), provides some support for the ab initio predictions. The evident failure of the semi-empirical calculations in this kind of problem is a consequence of the assumptions inherent in such methods; the neglect of electron repulsion integrals results in an artificial overweighting of nuclear-electron attractions in the more bonded, bridged structure (*12*). Thus, while semi-empirical calculations give generally reasonable results with problems not involving changes in connectivity (i.e., changes in the number of bonds), they should not be used to compare structures of differing connectivity.

IV. *n*-PROPYL AND ISOBUTYL CATIONS

Although the addition of a methyl group to the classical ethyl cation would formally destroy the sixfold rotational symmetry, it might be expected that the *n*-propyl cation would have only a low barrier around the CH_2^+–CH_2 bond. Ab initio calculations indicate this *not* to be the case (*20*). Instead, conformation II (which can be loosely described as the "C–C hyperconjugation" form) is found to differ from V (the "C–H hyperconjugation" form) by nearly 3 kcal/mole! Similar behavior is predicted for the isobutyl cation, VI–IX; the favored conformation (IX) is that with the smallest dihedral angles between the vacant p orbital at C_1 and the C–C bonds to the methyl groups.

These results are surprising for two reasons. Not only are the differences in energy among the various conformations, II–V and VI–IX, quite large, but the most stable forms, II and IX, are those which seem to indicate that "C–C hyperconjugation" is more effective than "C–H hyperconjugation." From work with substituted benzene derivatives, the reverse has long been thought to be true (21). In a rate of an equilibrium process, a 3 kcal/mole effect is equivalent to ~10^2. For this reason, it should be easy to detect a major conformational influence on reactivity. We believe we have found such an effect in the reactions of conformationally locked, bridgehead systems.

V. CONFORMATIONAL ANALYSIS OF BRIDGEHEAD REACTIVITY

Bartlett first realized the practical consequences of preferred planarity of carbonium ions: bridgehead carbonium ion reactions should be strongly inhibited (2). In more recent years, a wide range (over 10^{12}) in bridgehead solvolytic reactivity has been found (2). As expected, the larger, less restricted ring systems were the most reactive, since the carbonium ion intermediates could most nearly achieve planarity. However, the matter is not simple. Both 1-adamantyl bromide (X) and 1-bicyclo[2.2.2]octyl bromide (XI) have virtually the same tetrahedral geometry around the reaction site, and yet differ in reactivity by more than 10^3.

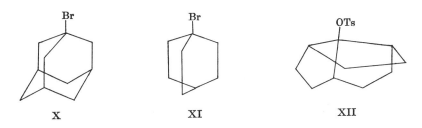

We have used computer conformational analysis to put bridgehead reactivity on a quantitative, predictable basis (10). In the Westheimer-Hendrickson-Wiberg-Allinger approach, various terms—angle strain, torsional strain, bond stretching strain, nonbonded interactions, etc.—are evaluated in a classical mechanical force field (9). We have used a modified Wiberg program, one that automatically seeks geometries of lowest energy. Force constants, empirically derived, have been included to represent carbonium ions.

The program works in the following way: The operator locates the coordinates for the carbon atoms for the system of interest and the computer program does the rest, from adding the necessary hydrogens to calculating the minimized energies. Two such energies are needed. The first, for the parent hydrocarbon, approximates the ground-state energy. The second calculation is performed on the carbonium ion, which is taken as a model for the solvolysis transition state. What really is being estimated is the enthalpy difference between hydrocarbon and carbonium ion; entropy and solvation effects are ignored. It is hoped that for a similar series of compounds these simplifications do not introduce serious errors. Furthermore, there are known errors in the strain energies estimated for some of the polycyclic hydrocarbons. Such errors tend to cancel by subtraction, when the cation–hydrocarbon energy differences

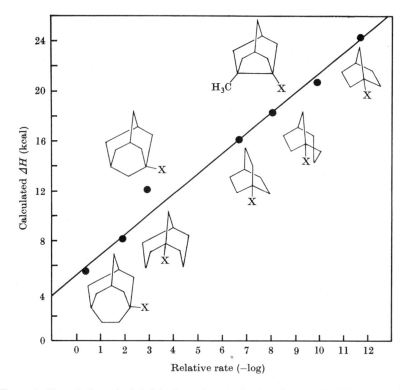

Figure 1 Plot of the calculated hydrocarbon–carbonium ion strain differences (ΔH) against −log of the experimental relative solvolysis rate constants.

are calculated. It is these energy differences which are plotted against the experimentally determined values, $\log k$ (solvolysis).

Despite the simplifications, approximations and inherent defects, the method works astonishingly well (*10*) (Fig. 1 is illustrative). The average deviation for the seven compounds is only $10^{\pm 0.5}$—a factor of 3—out of a total rate spread of nearly 10^{12}! Additional compounds, studied more recently, give equally good results (*22*). These afford examples of true predictions—the calculations were carried out before the rate constants were determined.

VI. DEVIATION OF 10-PERHYDROTRIQUINACYL BEHAVIOR

One compound, 10-perhydrotriquinacyl (10-tricyclo[5.2.1.0^{4,10}]decyl) tosylate (XII), deviates fantastically—by 10^{11}!—from the calculated value. This deviation, far outside the error limits of the calculation method, sugggests that some special effect may be responsible. Examination of models shows that XII is cup-shaped, with rear-side solvation at least a theoretical possibility. The other bridgehead derivatives examined (X, XI and the compounds in Fig. 1) have "cage" structures blocking such rear-side solvation. However, if solvation were a factor in the behavior of XII, it should react *faster* than predicted rather than *slower*, as is actually found.

Another structural difference between XII and the remainder of the bridgehead derivatives examined is to be found in the conformational arrangement around the reaction site. In XII the three adjacent CH bonds are *eclipsed* with the carbon-leaving group bond, but in X and XI, the three α-CH$_2$ groups are *staggered* relative to the bond to the leaving group. This basic geometrical difference, we believe, may provide the explanation for the 10^{11} deviation, described above. The carbonium ion from XII can be approximated by VI, *the highest-energy isobutyl cation conformation*. In contrast, the 1-adamantyl (from X) and 1-bicyclo[2.2.2]octyl (from XI) cations can be approximated by II, *the lowest-energy n-propyl cation conformation*. Relative to the cations from X and XI, the XII cation should thus be destabilized to a large degree, and this is consistent with the observed 10^{11} rate depression on solvolysis of XII.

The barriers calculated for the *n*-propyl and isobutyl cations are on the order of 3 kcal/mole. Since X–XII are tertiary, not primary, systems, this 3 kcal/mole effect must be multiplied by three because of the three equivalent arrangements around the reaction site. The result—9 kcal/mole, or about 10^7— is not quite large enough to account for all the observed effect. However, the *n*-propyl and isobutyl cation calculations were performed on planar cation structures. In the bridgehead systems, X–XII, the cations presumably are not planar. For this reason we have repeated the *n*-propyl cation calculations assuming a tetrahedral-like geometry of the C–CH$_2^+$ group. Such a nonplanar *n*-propyl cation is indicated to have a larger barrier—about 6 kcal/mole. The most stable conformation found (XIII) approximates the structures present in the nonplanar cations from X and XI. Structure XIV, the arrangement

XIII   XIV

expected from XII, is particularly unfavorable. Thus, the enhancement of the energy difference due to nonplanarity should serve to account for the 10^{11} rate difference between the solvolysis rate of XII and the other bridgehead systems exemplified by X and XI.

These ideas must, of course, be subjected to further experimental test. Theory is most effective when used in conjunction with experiment. It seems, however, that we have reached the point where theory effectively can *lead* experiment in suggesting where important and unexpected problems may lie.

ACKNOWLEDGMENTS

Finally, I should like to thank my many associates and students who contributed to this work. The quantum mechanical calculations were carried out in collaboration with Professor J. Pople of Carnegie-Mellon University, and with my colleague at Princeton, Professor L.C. Allen. The work was supported from grants provided by the National Science Foundation, the National Institutes of Health, and the Petroleum Research Fund, administered by the American Chemical Society. I would like to express my appreciation to the organizers of this conference for affording me the opportunity of presenting this paper.

REFERENCES

1. D. Bethell and V. Gold, "Carbonium Ions, An Introduction," Academic Press, New York, 1967.
2. R.C. Fort, Jr., and P.v.R. Schleyer, "Advances in Alicyclic Chemistry," Vol. 1, Academic Press, New York, 1966.
3. G.A. Olah and A.M. White, *J. Am. Chem. Soc.* **91**, 6883 (1969) and previous papers in the same series.
4. H. Hogeveen and C.J. Gaasbeek, *Rec. Trav. Chim.* **88**, 1305 (1969) and previous papers in the same series.
5. G.A. Olah, J.R. DeMember, C.Y. Lui and A.M. White, *J. Am. Chem. Soc.* **91**, 2958 (1969); A. Commeyras and G.A. Olah, *J. Am. Chem. Soc.* **91**, 2929 (1969).
6. G.A. Olah, A. Commeyras, J.R. DeMember and J.L. Bribes, *J. Am. Chem. Soc.*, in press.
7. L.N. Ferguson, *J. Chem. Ed.* **47**, 46 (1970).
8. R.J. Gillespie, *J. Chem. Ed.* **47**, 18 (1970).
9. J.E. Williams, P.J. Stang and P.v.R. Schleyer, *Ann. Rev. Phys. Chem.* **19**, 531 (1968).
10. G.J. Gleicher and P.v.R. Schleyer, *J. Am. Chem. Soc.* **89**, 582 (1967); P.v.R. Schleyer, P.R. Isele and R.C. Bingham, *J. Org. Chem.* **33**, 1239 (1968); unpublished results.
11. J.E. Williams, Jr., R. Sustmann, L.C. Allen and P.v.R. Schleyer, *J. Am. Chem. Soc.* **91**, 1037 (1969).
12. R. Sustmann, J.E. Williams, M.J.S. Dewar, L.C. Allen and P.v.R. Schleyer, *J. Am. Chem. Soc.* **91**, 5350 (1969).
13. J.E. Williams, Jr., V. Buss, L.C. Allen, P.v.R. Schleyer, W.A. Lathan, W.J. Hehre and J.A. Pople, *J. Am. Chem. Soc.* **92**, 2141 (1970).

14. P.G. Gassman, J.M. Hornback and J.L. Marshall, *J. Am. Chem. Soc.* **90**, 6238 (1968);
     F.B. Miles, *J. Am. Chem. Soc.* **90**, 1265 (1968).
15. V. Buss and L.C. Allen, unpublished calculations.
16. M.J.S. Dewar, "The Molecular Orbital Theory of Organic Chemistry," McGraw-Hill,
     New York, 1969; H.H. Jaffe, *Accounts Chem. Res.* **2**, 136 (1969).
17. J.P. Lowe, *Progr. Phys. Org. Chem.* **6**, 1 (1968).
18. References 12, 13 and unpublished calculations.
19. P.C. Myhre and E. Evans, *J. Am. Chem. Soc.* **91**, 5641 (1969); P.C. Myhre and K.S.
     Brown, *J. Am. Chem. Soc.* **91**, 5639 (1969).
20. J.A. Pople, L. Radom and V. Buss, unpublished calculations.
21. M.J.S. Dewar, "Hyperconjugation," Ronald Press, New York, 1962.
22. R.C. Bingham, unpublished results.

Note added in proof. Since the submission of this paper, several pertinent theoretical papers have appeared:

23. G.V. Pfeiffer and J.G. Jewett, *J. Am. Chem. Soc.* **92**, 2143 (1970).
24. D.T. Clark and D.M.J. Lilley, *Chem. Comm.* **1970**, 549, 603.
25. H. Kollmar and H.O. Smith, *Tetrahedron Letters* **1970**, 1833; *Angew. Chem.* **82**, 444
     (1970).

Recent Applications of Nuclear Magnetic Resonance Spectrometry in Conformational Studies of Cyclohexane Derivatives

Robert D. Stolow
Department of Chemistry
Tufts University
Medford, Massachusetts

For monosubstituted cyclohexanes (I), in *every* valid case reported, the chair conformation with the substituent *equatorial* (Ie) predominates at equilibrium (*1, 2*). However, introduction of a symmetrical polar transannular substituent,

$$(e) \qquad\qquad (a)$$
$$\Delta G^\circ = -RT \ln K$$

I $Z = H_2C$

II $Z = Y_2C$, $O{=}C$, Y_2N^+, O, S, O_2S, etc.

III $Z = HYC$, $YY'C$, HYN^+, etc.

Z, to give II, usually changes the conformational free-energy difference, ΔG°, from what it was for I. In certain cases, the axial conformation (IIa) predominates (*3*). The transannular interaction between polar X and Z groups which influences the relative free energies of IIa and IIe may be electrostatic in nature, so that

$$\Delta G^\circ_{II} - \Delta G^\circ_{I} = \Delta G^\circ_{II \,(electrostatic)} \tag{1}$$

Similarly, one may consider the introduction of an unsymmetrical polar transannular substituent, Z, to give III. Here the influence of a transannular electrostatic interaction would appear as follows:

$$(\Delta G^{\circ}_{III} - \Delta G^{\circ}_{III(X=H)}) - \Delta G^{\circ}_{I} = \Delta G^{\circ}_{III \text{ (electrostatic)}} \qquad (2)^{*}$$

If equations (1) and (2) are valid, then by evaluation of conformational free-energy differences, electrostatic interaction energies, $\Delta G^{\circ}_{II(electrostatic)}$ and $\Delta G^{\circ}_{III \text{ (electrostatic)}}$, can be determined experimentally, providing a tool for use in elucidation of the mechanism of electrostatic interactions in saturated molecules. Chlorocyclohexanes with transannular polar substituents have been studied with this aim in view. A common method for the estimation of "dipole–dipole" interaction energies has been tested and has failed that test. Before presenting these results, some general comments on experimental methods are offered for consideration.

Nuclear magnetic resonance (nmr) methods of conformational analysis have been reported based upon (a) chemical shifts, (b) vicinal coupling constants, (c) band widths and (d) peak areas. Jensen recently commented upon each of these methods as applied to cyclohexanes (2). Although peak-area measurement requires low temperatures (usually −150 to −70°C),† it is the only method of the four which requires neither model compounds nor temperature-dependence extrapolations. For methods a–c, model compounds with conformation-holding groups added are often used to approximate the individual chair conformations, e and a. Difficulties in obtaining suitable model compounds are particularly acute for the chemical-shift method, since remote substituents often influence chemical shifts significantly. Chemical shifts of model compounds may also be temperature- (2), solvent- and concentration-dependent. Vicinal coupling constants and band widths of model compounds often appear to be less sensitive to these variables than are the chemical shifts. However, for those compounds where all four methods may be applied, the chemical shifts, coupling constants, and band widths of e and a may be obtained from spectra of the mixture of e and a‡ over a range of low temperatures, and the individual values for e and a may be extrapolated to the higher temperatures required by methods a–c (2). In such cases, a comparison of the results of all four methods would be of interest.

Remote deuterium substitution to remove long-range coupling is usually unnecessary when measuring peak areas. Remote deuterium substitution may

* By definition, for II, $\Delta G^{\circ}_{II \ (X=H)} \equiv 0$.

† Temperatures below −150°C (not readily accessible with commercial nmr spectrometers) would be required to obtain peak areas for cyclohexanones, and other compounds with relatively low barriers to conformational interconversion.

‡ Separation of e from a at low temperatures and observation of each individual conformation would be ideal.

alter slightly the relative conformational free energies, or independent of such effects, may change slightly the chemical shifts and coupling constants. Nevertheless, if methods b and c are to be used, the best results are obtained for cyclohexanes with no spin–spin couplings except those among the protons at C-1, C-2, and C-6. Then, accurate band widths may be obtained even when complete analysis of the vicinal couplings is not possible.

When peaks for e and a overlap at low temperatures, vicinal deuterium substitution may make peak-area measurements more accurate. However, cyclohexanes with vicinal deuterium are useless when vicinal coupling constants and band widths are desired (methods b and c).

Table I   Band Widths (W) and Chemical Shifts (δ) of 4-Chlorocyclohexanone-2,2,6,6-d_4 (IV) and Model Compounds V and VI[a]

Compound	δ (Hz)	W (Hz)
IV	444.74 ± 0.1	18.05 ± 0.05[b]
V	451.4 ± 0.1	11.38 ± 0.2
VI	431.4 ± 0.1	31.45 ± 0.1

[a] In carbon tetrachloride solution at 31°C and 100 MHz. Data given are for the C-4 protons.
[b] The value reported (4), $W = 19.4$ Hz for IV in carbon tetrachloride, may differ from this value in part because of differences in temperature and concentration.

Impurities usually interfere least with determinations of chemical shift and band width, but can interfere seriously with the complete analysis of a spectrum often necessary to extract accurate coupling constants. Impurities are most troublesome when they appear near or overlap peaks, the areas of which are to be determined. Therefore, highly purified samples are often required for peak-area measurements.

Peak-area and band-width studies have been found to be most useful in the present work. Many of the points made above may be illustrated by the results of nmr conformational studies of some derivatives of chlorocyclohexane.

The equatorial conformation (Ie) predominates for chlorocyclohexane (I, X = Cl), Ie $\rightleftharpoons$ Ia, $\Delta G° = 0.43$ kcal/mole at room temperature (1), and $\Delta G° = 0.53 \pm 0.02$ kcal/mole at $-81°C$ (2). However, for 4-chlorocyclohexanone-2,2,6,6-d_4 (IVe $\rightleftharpoons$ IVa) in carbon tetrachloride solution, the *axial conformation predominates* (Table I). By use of data from Table I, two independent values of the conformational free-energy difference for the equilibrium, IVe $\rightleftharpoons$ IVa, may be determined and compared, one by use of the chemical-shift method (equations 3 and 5) and the other by use of the band-width method (equations 4 and 5).

$$N_{IVa} = (\delta_{VI} - \delta_{IV})/(\delta_{VI} - \delta_{V}) \tag{3}$$

$$N_{IVa} = (W_{VI} - W_{IV})/(W_{VI} - W_{V}) \tag{4}$$

$$\Delta G°_{IV} = -RT \ln [N_{IVa}/(1 - N_{IVa})] \tag{5}$$

Both methods give the same result when 2,6-dimethyl-4-chlorocyclohexanones (V and VI) are used as model compounds for IV: $67 \pm 1\%$ IVa; $\Delta G°_{31°C} = -0.42 \pm 0.03$ kcal/mole.* It would be of interest to extend this comparison to other solvents and temperatures. However, more important is the nature of the significant influence of the carbonyl group upon the relative conformational energies: $\Delta\Delta G° = 0.43 - (-0.42) = 0.85$ kcal/mole. Most of this difference apparently can be accounted for by the classical method of Jeans for calculation of dipole–dipole interaction energies (5), if the dielectric constant of the medium between the dipoles (D) is treated as an adjustable parameter different for each conformer (6). This procedure seemed rather arbitrary, and a critical test of the method of Jeans appeared warranted.

---

* The limits of error do not include the approximation of using V and VI as models for IVa and IVe, respectively. Other sets of models should also be considered. However, V and VI may be the best available set. If VIIe and VIIa (below) are used as band-width models with no correction for solvent difference and temperature (fluorotrichloromethane at $-74°C$), then $\Delta G°$ would be -0.48 kcal/mole, in good agreement with -0.42 kcal/mole. But VIIe and VIIa at $-74°C$ in fluorotrichloromethane are useless as models for the chemical-shift method since both are upfield from IV.

For the ethylene ketal of IV, 8-chloro-1,4-dioxaspiro[4.5]decane-2,2,3,3,6,6,10,10-d_8 (VII), the approximate method of Jeans (5) predicts relative destabilization of VIIa by the ketal group regardless of the values of D chosen (when $D \geqslant 1$). The results presented below are in opposition to the prediction of the method of Jeans. Rather than consider fractional or negative values of D, it seems preferable to conclude that *the approximate method of Jeans does not work.*

The C-8 proton of 8-chloro-1,4-dioxaspiro[4.5]decane-2,2,3,3-d_4 (VIII) gives a very complex nmr multiplet at 31°C in 1 *M* solution in fluorotrichloromethane. It was not possible to determine accurately the chemical-shift

VII(e) A = D
VIII(e) A = H

VII(a) A = D
VIII(a) A = H

position or the band width for the C-8 proton of VIII. Determination of vicinal coupling constants for this AA'BB'CC'DD'X spin system would be a hopeless task. Therefore the d_8 derivative (VII) was prepared for this work, so that chemical shifts, band widths, and vicinal coupling constants for the resulting AA'BB'X spin system might be determined, as well as peak areas. The C-8 proton of VII, 1 *M* in fluorotrichloromethane, gave, at 31°C, a nine-line multiplet (1:2:2:1:4:1:2:2:1 approximate intensity ratios) composed of sixteen transitions, as expected (7). The distance between the outer lines of the C-8 proton multiplet (sharp peaks, each resulting from single transitions) gave an accurate band width, $W = 21.54 \pm 0.06$ Hz, for VII under these conditions. At −40°C, 1 *M* solutions of VII or VIII in fluorotrichloromethane showed considerable broadening of the C-8 proton multiplet. At −74°C, the rate of conformational interconversion was slow on the nmr time scale, and sharp peaks were observed for individual chair conformations, separated by 74.5 Hz at 100 MHz. Electronic integration of peak areas showed $58.3 \pm 2\%$ of VIIa (average of 45 scans) as compared to $58.4 \pm 2\%$ of VIIIa (average of 51 scans), under the same conditions. Apparently, the presence of deuterium on the cyclohexane ring in VII has no significant effect upon the relative conformational free energies as compared to VIII.* For both VIIe $\rightleftharpoons$ VIIa and VIIIe $\rightleftharpoons$ VIIIa, $\Delta G^\circ_{-74^\circ C} = -0.13 \pm 0.03$ kcal/mole. Using the value for chlorocyclohexane, $\Delta G^\circ_{-81^\circ C} = 0.53 \pm 0.02$ kcal/mole (2), then at $-74^\circ C$ $\Delta\Delta G^\circ = 0.53 - (-0.13) = 0.66$ kcal/mole, a substantial relative stabilization of the *axial* conformation, VIIa or VIIIa, rather than the destabilization predicted by the approximate method of Jeans (5, 6).

At $-74^\circ C$, the band widths for VIIa, 11.65 ± 0.05 Hz, and for VIIe, 32.32 ± 0.05 Hz, are clearly detectable. The temperature-dependence of the band widths of VII in fluorotrichloromethane and in acetone-d_6 solutions are shown in Table II.

Table II   Temperature-Dependence of Band Widths of 8 - Chloro - 1,4 - dioxaspiro[4.5]decane - 2,2,3,3,6,6,10,10-d_8 (VII)[a]

Solvent	T (°C)	W (Hz)	ΔG° (cal/mole)
$(CD_3)_2CO$	64	22.87 ± 0.05	$+110 \pm 10$
	31	22.90 ± 0.05	$+100 \pm 10$
	-18	22.83 ± 0.11	$+ 90 \pm 20$
	-74	32.45 ± 0.05^b	
		11.55 ± 0.05^c	
$FCCl_3$	31	21.54 ± 0.06	$- 50 \pm 10$
	-18	21.12 ± 0.10	$- 80 \pm 20$
	-74	32.32 ± 0.05^b	
		11.65 ± 0.05^c	

[a] In fluorotrichloromethane and in acetone-d_6 solutions. Band widths given are for the C-8 proton.
[b] For VIIe.
[c] For VIIa.

The solvent-dependence of the band width of VII is shown in Table III. The more polar solvents, acetone-d_6 and dimethyl sulfoxide-d_6, tend to stabilize the equatorial conformation, VIIe, relative to VIIa, as compared to the less polar solvents, carbon tetrachloride and fluorotrichloromethane. Of the four solvents, the spectrum of VII at 31°C in dimethyl sulfoxide-d_6 (Table III) appeared to show the largest separation of the chemical shifts of the geminal

* Unlike the observation of Jensen (2), change in irradiation power level (all below saturation) had no significant influence upon the observed peak-area ratios.

Table III   Solvent-Dependence of the C-8
Proton Band Width of VII[a]

Solvent	W (Hz)	$\Delta G°$ (cal/mole)[b]
CCl_4	21.18 ± 0.05	-90 ± 20
$FCCl_3$	21.54 ± 0.06	-50 ± 10
$(CD_3)_2CO$	22.90 ± 0.05	$+100 \pm 10$
$(CD_3)_2SO$	23.02 ± 0.07	$+120 \pm 20$

[a] At 31°C.
[b] Based upon −74°C band widths for VIIe and VIIa, Table II.

protons (A and B) of the AA'BB'X spin system. Analysis of the spectrum by use of the LAOCN 3 program gave the following results:

$$J_{AB} = -13.43 \, \text{Hz} \qquad \delta_A - \delta_B = 18.5 \, \text{Hz}$$

$$J_{AX} = 7.82 \pm 0.02 \, \text{Hz (splitting 7.46 Hz)}$$

$$J_{BX} = 3.64 \pm 0.02 \, \text{Hz (splitting 4.04 Hz)}$$

$$W_{calc} = 2(J_{AX} + J_{BX}) = 22.92 \pm 0.08 \, \text{Hz}$$

$$W_{observed} = 23.02 \pm 0.07 \, \text{Hz}$$

Because $\delta_A - \delta_B$ is only slightly greater than J_{AB} in absolute magnitude, the observed splittings are closer to one another than the actual coupling constants. In such cases, splittings cannot be substituted for couplings for use in the coupling-constant method of conformational analysis. For VII, if J_{AX} could be extracted from the −74°C spectrum for each component of the equilibrium mixture of VIIa and VIIe, the coupling-constant method could then be applied properly.* Triple resonance experiments at −74°C would probably permit the overlapping lines in the AB region to be assigned and the J_{AX} values to be extracted for VIIa and VIIe. Experimentally, the peak-area and band-width studies are more readily carried out.

Under all conditions studied, the chloroketal (VII) gives results opposed to the prediction of the method of Jeans (5). The ketal group of VII or VIII, like the carbonyl group of IV, stabilizes *axial* relative to equatorial chlorine. Because it has failed this critical test, *use of the approximate method of Jeans for evaluation of intramolecular electrostatic interactions must be abandoned.*

* For VIIa and VIIe, J_{BX} values are expected to be too close to one another to be useful in conformational studies.

Comparing ketone IV and ketal VII, although the carbonyl dipole and the ketal group dipole have opposite directions, their transannular electrostatic interactions have the same sign:

$$\Delta G^{\circ}_{\text{IV(electrostatic)}} = -0.85 \text{ kcal/mole} \qquad (CCl_4, 31°C)$$

$$\Delta G^{\circ}_{\text{VII(electrostatic)}} = -0.48 \text{ kcal/mole} \qquad (FCCl_3, 31°C)$$

Apparently, the direction of the group dipole is not the controlling factor, but the fact that both the carbonyl and ketal groups induce a partial positive charge at Z (structure II) may be most significant.* The importance of the charge at Z (structure II) deserves further consideration.

A unit positive charge can be placed at the transannular position in 4-chloro-N,N-dimethylpiperidinium ions. Again, the results show relative stabilization of the conformation with chlorine axial, IIa (Z = $(CH_3)_2N^+$) (8, 9).

Attempts at elucidation of the mechanism of electrostatic interaction in these and other examples would best focus attention upon the microscopic properties of the system, avoiding such macroscopic ideas as effective dielectric constant of the medium between two points within a molecule. Fresh approaches to the problem are needed.

ACKNOWLEDGMENTS

I would like to acknowledge the assistance of T.W. Giants and D.I. Lewis in the work described in the present paper. I wish to thank the National Science Foundation and the Research Corporation for their support.

REFERENCES

1. J.A. Hirsch, "Table of Conformational Energies—1967," in "Topics in Stereochemistry," Vol. 1, N.L. Allinger and E.L. Eliel, eds., Wiley (Interscience), New York, 1967.
2. F.R. Jensen, C.H. Bushweller and B.H. Beck, J. Am. Chem. Soc. 91, 344 (1969).
3. R.D. Stolow, T.W. Giants and J.D. Roberts, Tetrahedron Letters 1968, 5777; R.D. Stolow, T. Groom and P.D. McMaster, Tetrahedron Letters 1968, 5781.
4. E. Premuzic and L.W. Reeves, Can. J. Chem. 42, 1498 (1964).
5. E.L. Eliel, N.L. Allinger, S.J. Angyal and G.A. Morrison, "Conformational Analysis," pp. 460–462, Wiley (Interscience), New York, 1965.
6. T. Groom, Ph.D. Dissertation, Tufts University, 1969.
7. J.W. Emsley, J. Feeney and L.H. Sutcliffe, "High Resolution Nuclear Magnetic Resonance Spectroscopy," Vol. 1, pp. 416–423, Pergamon Press, Oxford, 1965.
8. M.L. Stien, G. Chiurdoglu, R. Ottinger, J. Reisse and H. Christol, Tetrahedron 26 (1970), in press.
9. R.D. Stolow, D.I. Lewis and P.A. D'Angelo, Tetrahedron 26 (1970), in press.

* Other Compounds of type II with Z = Cl_2C and F_2C behave similarly (3).

Conformational Aspects of the Reaction of the 1-Methylcyclodecane-1,6-diols with Acid

H.H. Westen
Organisch-chemisches Laboratorium
Eidgenössische Technische Hochschule
Zurich, Switzerland

Conformational analysis in the cyclodecane series was given a firm basis by Dunitz and co-workers, who determined the crystal structures of several cyclodecane derivatives by x-ray analysis and found the carbon ring to have conformation I, except when the substitution pattern was incompatible with it (*1*). Since much of the chemistry peculiar to the medium-sized rings had

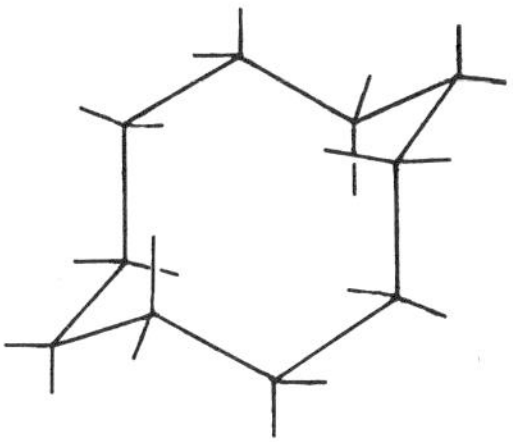

I

Scheme I

been done earlier, it seemed appropriate to reevaluate the results in the light of the conformational model I.

A series of reactions published by Prelog and Küng in 1956 (*2*) is of particular interest with respect to the conformational requirements for transannular reactions. They reported that 6-hydroxycyclodecanone is converted by methylmagnesium iodide in refluxing benzene to a *single* product in over 80% yield; it was characterized as 1-methylcyclodecane-1,6-diol and assumed to be one of the two possible diastereoisomers. This compound, when treated with 84%

259

phosphoric acid at room temperature, yielded, in addition to other products, 6-methylcyclodecanone as a result of a reaction that involved a transannular hydride shift. By labeling the educt with deuterium at the secondary carbinol carbon and converting the product to 5-methylsebacic acid—which still contained all the deuterium—it was shown that the hydrogen had migrated across the ring from the labeled position. A 1,6 shift seemed the obvious conclusion. In the light of conformation I, however, one might consider a 1,2 shift followed by a 1,5 shift as a possible alternative, which had not been ruled out by the experiment. In a reinvestigation of this work, additional products were found. The experimental results will therefore be described before their conformational implications will be discussed.

The Grignard reaction was carried out with some modifications. As it cannot be effected in ether (2) and as benzene presents problems of solubility, boiling tetrahydrofuran was chosen as the solvent, and methylmagnesium bromide was used instead of the iodide. With 3.3 moles of the Grignard reagent per mole of 6-hydroxycyclodecanone and 3 hr reaction time, 1-methylcyclodecane-1,6-diol was obtained in 70% yield; 27% of starting material was recovered. After recrystallization from ethyl acetate, the product had indeed a sharp melting point at 128.5–129°C, which remained unchanged after further recrystallizations, confirming the result of Prelog and Küng (2). Chromatography of the residue from the mother liquor with ethyl acetate on silica gel, however, afforded, beside the unreacted starting material, two more crystalline substances in roughly equal amounts, which turned out to be the expected diastereoisomers. The less polar one had mp 137–137.5°C, the other one crystallized in two modifications—sometimes side by side from the same solution: fine needles with mp 123.5–124°C and platelets or flat prisms with mp 124.5°C. Crystals of either modification could be grown by recrystallizing either one with proper seeding. Perhaps the two modifications represent crystals of the two possible conformations II and III (Scheme III) with a relative configuration of the secondary hydroxyl group as yet unknown, just as *trans*-1,6-diaminocyclodecane dihydrochloride can be obtained in two crystalline modifications differing in the conformational arrangement of the substituents (1). More important, when a 1:1 mixture of the two diastereoisomers was recrystallized from ethyl acetate, crystals formed that had the same characteristic appearance and melting point as those obtained by recrystallizing the crude product. As the infrared spectrum of this 1:1 crystallizate in nujol is different from that of either of the diastereoisomers and in particular lacks at least one of the strong bands of the less polar isomer, the 1:1 crystallizate cannot be a solid solution.

1-Methylcyclodecane-1,6-diol was then labeled at C-6 by oxidation with silver carbonate on celite in refluxing benzene (3) and reduction of the resulting ketone by lithium aluminum deuteride, and was finally separated into its diastereoisomers. Treatment of each isomer with 84% phosphoric acid resulted in product mixtures of identical composition, from which 6-methylcyclodecanone was isolated by chromatography and examined by nuclear magnetic resonance (nmr). The methyl protons of each sample gave rise to a broad singlet, which sharpened upon decoupling* of the deuterium thus leaving no doubt as to the absence of any noticeable amount of molecules with ^{1}H at C-6 and confirming a pure 1,6 shift in the reaction.

The olefin fraction of the product can be separated by gas chromatography into two major components in approximately equal amounts and a third one of very low intensity. The nmr spectrum of the mixture shows that there is one

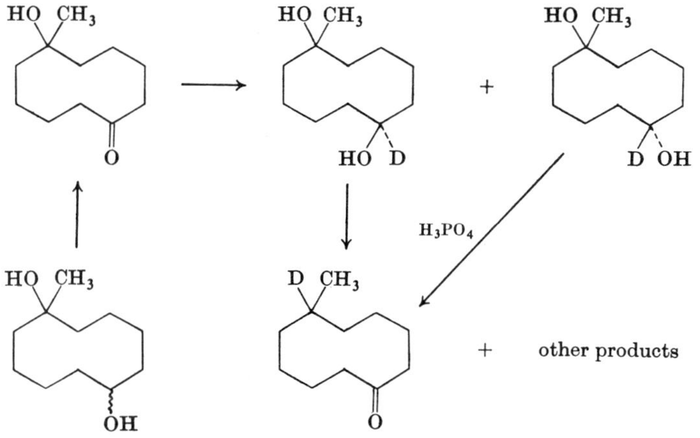

Scheme II

methyl group in a vinylic position and another geminal to a hydrogen in an allylic position, but not more than traces of vinylic hydrogens. These results indicate the structures IV and V (Scheme III), the alternative of an octaline skeleton having been excluded by Prelog and Küng (2), who hydrogenated the mixture and found the infrared spectrum of the product to be different from that of any of the methyldecalins.

In addition to the olefins IV and V (19%) and the ketone VI (42%), two alcohols were isolated, one of which (14.5%) proved to be identical with authentic samples (4, 5) of *trans,trans*-9-methyl-1-decalol (VII) by comparison of the infrared spectra (4) and by melting and mixed melting points of the 3,5-dinitrobenzoates (5). The other one (6.5%), according to infrared, nmr and mass spectral data, appears to be 6-methyl-5-cyclodecen-1-ol (VIII), probably the *cis* isomer.

* The decoupling experiment was performed by Mr. W. Schittenhelm of Spectrospin AG, Zürich, whose help is gratefully acknowledged.

The course of the reactions involved can be summarized as in Scheme III. With the restriction that geminal substituents can only be located at carbon atoms of the conformational type II (*1*) in the stable conformation of the ten-membered ring, each of the two diastereoisomers of 1-methylcyclodecane-1,6-diol can adopt two favorable conformations: the methyl group being either semi-equatorial (*1*) as in II or semi-axial (*1*) as in III. When the tertiary

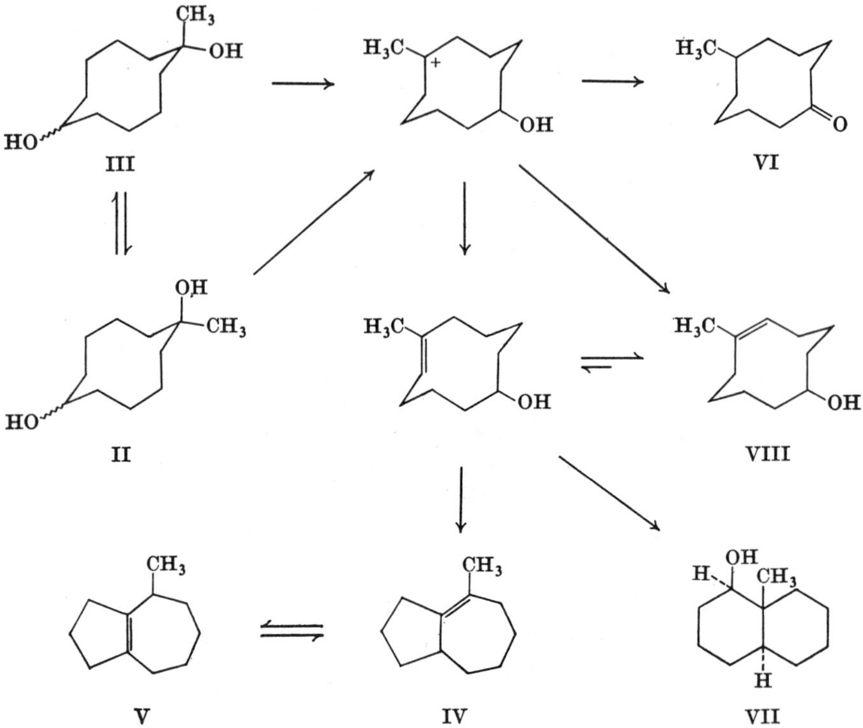

Scheme III

hydroxyl is eliminated, this difference is preserved in the corresponding conformation of the secondary hydroxyl, whereas the diastereoisomeric difference is lost. Since the product distribution is the same for the two diastereoisomers, the transannular hydride shift in the formation of 6-methylcyclodecanone does not seem to be concerted with the elimination of the hydroxyl group. On the contrary and as expected, the tertiary carbonium ion equilibrates conformationally before the reaction proceeds. Its trigonal carbon atom is believed to be mainly at one of the two antiperiplanar bonds, because much of the trans-annular strain is thus relieved. For the olefin formation, two of the vicinal hydrogens are then suitably arranged to be eliminated, the loss of one giving rise to a *cis*, and of the other to a *trans* double bond. Before the proton is removed, it can conceivably form a bridged ion, which in the *trans* case might have been prone to an attack by a hydride ion migrating 1,5 from the carbinol carbon. This possibility, which was the reason for the reinvestigation of the reaction, is, however, not realized, as shown by the results described above. A conformation of the carbonium ion similar to I does not seem to be favorable for a 1,6 hydride shift; it would lead to considerable crowding of both the methyl and the hydroxyl groups with intraannular hydrogens in the transition state. Therefore, conformations like IX would be better suited for this transannular reaction. However, this matter cannot be decided on the basis of our present knowledge and further work towards the elucidation of conformational requirements for transannular hydride shifts is in progress.

Under the conditions of the reaction, *trans*-6-methyl-5-cyclodecen-1-ol easily loses its secondary hydroxyl group, because the double bond is in a good position to participate with development of a transannular bond. If the

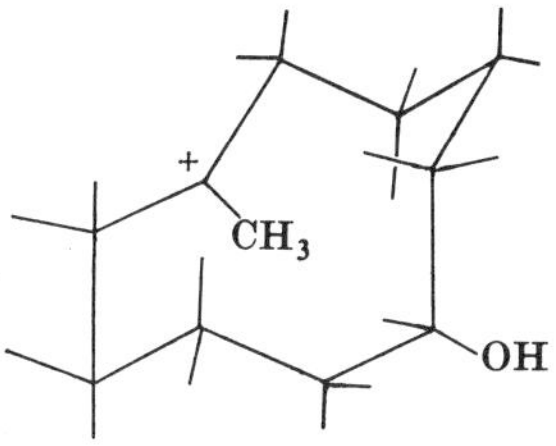

IX

Scheme IV

bridging occurs 1,5, a tertiary carbonium ion results, which is stabilized by loss of a proton, and 2-methylbicyclo[5.3.0]decene is the product, the double bond being equilibrated between the two ditertiary positions by acid catalysis. Whereas we cannot draw any conclusion as to the best conformation for this cyclization from the experimental results, because the stereochemical information is lost in the olefin formation, the product of the 1,6 bridging preserves such information. The latter cyclization is accompanied by solvation of the developing secondary carbonium ion at C-5, which is only accessible from the outside of the molecule. In a stereospecific reaction one thus obtains *trans,trans*-9-methyl-1-decalol, in which the relative configuration of the methyl to the hydroxyl group betrays the *trans* olefin as its precursor and the *trans* ring junction is evidence for a transition-state conformation similar to I rather than IX.

The stereochemical course of the decalol formation parallels that of the solvolysis of *trans*-5-cyclodecen-1-yl-*p*-nitrobenzoate investigated by Goering and Closson (*6*), who also solvolyzed the *cis* isomer and obtained *cis,cis*-1-decalol with a reaction rate over 300 times slower than that of the *trans* isomer. Our 9-methyl-1-decalol contained less than 1% of a compound other than the *trans,trans* isomer. Again, conformational analysis offers a satisfactory explanation: The best conformation of *cis*-6-methyl-5-cyclodecen-1-ol or the carbonium ion derived from it is not well-suited for cyclization.

Finally, the Grignard reaction deserves some comment. Among the four possible conformers X–XIII of 6-hydroxycyclodecanone having a ring conformation similar to I, XI will probably predominate in the equilibrium mixture, concluding, e.g., from data obtained by x-ray analyses of 1,6-cyclodecanedione (*1*) and 2-oxa-1,6-cyclodecanedione (*1*), in which the carbonyl groups prefer positions at bonds with antiperiplanar conformations. But XI as well as X would lead to transition states, in which the oxygen on the tertiary carbon would be forced towards intraannular hydrogens with considerable increase in transannular strain. On the other hand, XII and XIII

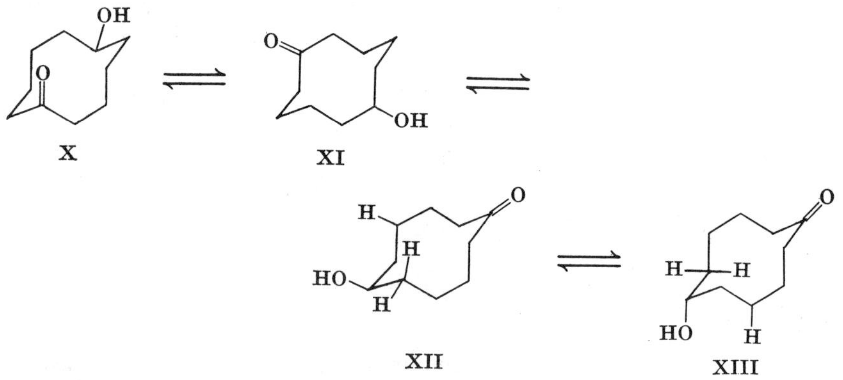

Scheme V

should add Grignard reagents without difficulty. They differ only by the conformational positions of their hydroxyl groups, one being semi-equatorial (*1*) and the other semi-axial (*1*). Substituents in these two positions are in similar environments. Conformers XII and XIII should therefore be energetically almost equivalent although much less favorable than XI, in which transannular strain is appreciably smaller. For these reasons, the reaction of 6-hydroxycyclodecanone with Grignard reagent may be expected to be relatively slow and to yield comparable amounts of the two diastereoisomeric products, as has been confirmed by the experiment.

From these examples, one can learn that conformational analysis of the ten-membered ring has reached a stage that allows us to understand *post festum* a great deal of the mechanistic and steric aspects of the pertinent chemistry. But there are still not enough data for even qualitatively correct predictions in every case.

REFERENCES

1. J.D. Dunitz, *in* "Perspectives in Structural Chemistry," Vol. II, p. 1, J.D. Dunitz and J.A. Ibers, eds.,Wiley, New York, 1968.
2. V. Prelog and W. Küng, *Helv. Chim. Acta* **39**, 1394 (1956).
3. M. Fetizon and M. Golfier, *Compt. Rend.* **267 C**, 900 (1968).
4. W. Acklin, V. Prelog, F. Schenker, B. Serdarević and P. Walter, *Helv. Chim. Acta* **48**, 1725 (1965).
5. C.B.C. Boyce and J.S. Whitehurst, *J. Chem. Soc.* **1960**, 2680.
6. H.L. Goering and W.D. Closson, *J. Am. Chem. Soc.* **83**, 3511 (1961).

Author Index

Numbers in parentheses are reference numbers and indicate that an author's work is referred to although his name is not cited in the text. Numbers in italics show the page on which the complete reference is listed.

A

Abdullaev, N. D., 111(3, 4), *128*
Abe, A., 209(11), *218*
Abraham, R. J., 95, *95*
Acklin, W., 261(4), *265*
Adams, W. J., 4(17a, 17b), *13*
Aglietto, M., 209(10), *218*
Allegra, G., 216(16, 17), *218*
Allen, G., 94(2), *108*
Allen, L. C., 133(36, 38, 53), 134(36, 37, 95), 139(108), 140(107), 141(119), 142 (108), 145(36, 37, 38), 146(36, 37), 146, 147(38), 147, 148(36, 37, 38, 44), 149 (44), 152(38, 107, 126), *153*, *154*, *155*, 242(11, 12, 13, 15), 243(12), 244(12), *248*, *249*
Allen, T. L., 226, *228*
Allinger, N. L., 2(11), 3(11), 4(11), 5(11, 21), 6, 9(21), *13*, 15(1), *28*, 65(4), 67(4), *71*, 100(6), 103(6), *108*, 157(1), *164*, 165(1), 167(3), 173(1), *175*, 191(3, 4), *202*, 223(14, 16), 225, 227, *228*, 251(1), 254(1, 5), 255(5), 256(5), 257(5), *258*
Altona, C., 1(1, 5, 6, 7), 2(5, 10, 11, 12), 3 (10, 11), 4(5, 6, 7, 10, 10a, 10b, 11, 14, 17, 29), 5(11, 17), 6(1, 23, 24, 29), 7 (10, 23), 8(24), 9(10, 24), 10(29), *13*, 62(6), *62*
Anderson, J. E., 16(6), 19(6), *29*
André, J. M., 152(123), *155*
André, M.-C., 152(123), *155*
Anet, F. A. L., 15(3), 16(3), 17(3), *28*, 16 (5), 17(5), 20(10), 23(10), *29*, 137(70), 140(85), *154*

Angyal, S. J., 15(1), *28*, 165(1), 173(1), *175*, 191(3, 4), *202*, 223(14), *228*, 254(5), 255(5), 256(5), 257(5), *258*
Anteunis, M., 32(2), 33(6, 7), 35(8, 9, 10, 11), 36(11, 12), 37(11), 42(13), *42*, 43 (14), 44(15, 16, 17, 18), 46(19, 22), 47 (2, 8, 9, 11, 12, 13, 15, 21), 49(21), *50*
Antonov, V. K., 111(2), 126(1), *127*
Arkhipova, S. F., 111(4), 120(11), *128*
Armstrong, D. R., 140(46), 141(46), 149 (49), *153*
Aronova, N. L., 163(9), *164*

B

Baird, N. C., 130(10, 10b), *152*
Baisted, D. J., 59(2), *62*
Bak, B., 135(5), 140(5), 142(5), 145(5), *153*
Barfield, M., 103(11), *109*
Barnett, M. P., 130(4), *152*
Barr, S. J., 2(8), *13*
Bartell, L. S., 4(17a, 17b), 5(19), *13*
Bartlett, N., 187(20), *189*
Bartolo, H. F., 219(4), *228*
Basch, H., 132(18), 140(80), *153*, *154*
Bautista, R. M., 137(69), *154*
Beck, B. H., 219(2), 220(2), 221(12), 222, 223(2), *228*, 251(2), 252(2,) 254(2), 256(2), *258*
Beckett, C. W., 220, *228*
Beecker, J. F., 223(19), *228*
Benedetti, E., 209(12), *218*
Benedict, W. S., 142(89), *154*
Bentley, F. F., 224(21), *228*

Subject Index

A

B

C

THE LIBRARY
UNIVERSITY OF CALIFORNIA
San Francisco

THIS BOOK IS DUE ON THE LAST DATE ording to the Library

Books not returned on time are subject to the Library Lending Code.

Books not in demand may be renewed if application is made before expiration of loan period.

5 - 6 - 71

14 DAY

14 DAY

JUL 3 1980 RETURNED

MAY 26 1971

JUN 20 1980

RETURNED

14 DAY

MAY 26 1971

JUL - 9 1981

14 DAY

OCT 18 1976 RETURNED

JUL - 8 1981

RETURNED

OCT - 7 1976

14 DAY

OCT 28 1976

RETURNED

OCT 21 1976

15m-5,'70(N6489s4)4128—A33-9